MATRIX

STANLEY M. DAVIS

Columbia University
Boston University

PAUL R. LAWRENCE

Harvard University

in collaboration with
Harvey Kolodny
University of Toronto
and
Michael Beer
Harvard University

ADDISON-WESLEY PUBLISHING COMPANY
Reading, Massachusetts • Menlo Park, California • New York
Don Mills, Ontario • Wokingham, England • Amsterdam • Bonn
Sydney • Singapore • Tokyo • Madrid • San Juan

This book is in the Addison-Wesley Series:

ORGANIZATION DEVELOPMENT

CONSULTING EDITORS:

Edgar H. Schein

Warren G. Bennis

Richard Beckhard

ISBN 0-201-01115-8

17 18 19 20-CRW-9998979695

FOREWORD

About two hundred years ago something began happening in England and Western Europe that literally changed the face of the earth and permanently altered humanity's place on it. Our school books sum it all up as the "Industrial Revolution," and in the popular view that phenomenon is associated largely with the invention of machines and the discovery of new forms of power. The result was an unprecedented and continuous increase in human productivity, in the ability to achieve even greater output from a given amount of labor and a given expenditure of capital. Many forget, however, that all this became possible only through the development of systems for organizing and controlling the machines, energy, and people in entirely new ways. In other words, the revolution was managerial as well as industrial and scientific.

These management techniques—nearly all of which grew out of ad hoc responses to particular problems of the moment, rather than from some theoretical design—gradually became institutionalized in the modern corporation. As sociologist Daniel Bell has observed:

> The church and the army have been the historic models of organizational life. The business corporation, which took its present shape in the first decades of the 20th century, was the one new social invention to be added to these historic forms.

Not surprisingly, early business structures derived many of their organizing principles from the historic military model, and most of them continue to do so as evidenced, for example, by the survival of such concepts as "staff" and "line" functions even within organizations in which the logical distinction can be maintained only with great ingenuity.

In this book, the authors invite us to consider a new alternative to what they call the "1-boss" command structure that evolved from the industrial revolution and which "is so close to being a universal pattern that managers have been unaware until recently that any choice existed in this matter." The matrix system they describe is one which grew out of the unique management problems of the American space effort of the 1960s. It comes to us with a strong caveat from the authors:

> If you do not really need it, leave it alone. There are easier ways to manage organizations.

As described in Chapter 8 of this book, Citicorp has been exploring the possibilities of matrix organization since 1973, and I can personally attest to the fact that there are, indeed, easier ways to manage. On the other hand, managers are paid to find, not the easiest way, but the best way. For a business, the best way is whatever coordinates people, materials, and money for the production of goods or services that customers want at the least cost with the best return on capital invested.

Matrix represents a sharp break with traditional forms of business organization designed to accomplish that objective. For those organizations which need such a system and can learn to use it fully, it offers stimulating possibilities. Others may wish to avoid it. The important point is that the spectrum of human institutions now offers us another choice in the selection of organization models. It should be carefully examined.

Walter B. Wriston
Chairman, Citicorp–Citibank

SERIES FOREWORD

The Addison–Wesley Series on Organization Development originated in the late 1960s when a number of us recognized that the rapidly growing field of "OD" was not well understood or well defined. We also recognized that there was no one OD philosophy, and hence one could not at that time write a textbook on the theory and practice of OD, but one could make clear what various practitioners were doing under that label. So the original six books launched what has since become a continuing enterprise, the essence of which was to allow different authors to speak for themselves instead of trying to summarize under one umbrella what was obviously a rapidly growing and highly diversified field.

By the early 1980s the series included nineteen titles. OD was growing by leaps and bounds, and it was expanded into all kinds of organizational areas and technologies of intervention. By this time, many textbooks existed as well that tried to capture the core concepts of the field, but we felt that diversity and innovation were still the more salient aspects of OD.

Now as we move into the 1990s our series includes twenty-seven titles, and we are beginning to see some real convergence in the underlying assumptions of OD. As we observe how different professionals working in different kinds of organizations and occupational communities make their case, we see we are still far from having a single "theory" of organizational development. Yet, a set of common assumptions is surfacing. We are beginning to see patterns in what works and what does not work, and we are becoming more articulate about these patterns. We are also seeing the field connecting to broader themes in the organizational sciences, and new theories and theories of practice are being presented in such areas as conflict resolution, group dynamics, and the process of change in relationship to culture. The new titles in the series address current themes directly: Tjosvold's *The Conflict-Positive Organization*, for example, connects to a whole research tradition on the dynamics of collaboration, competition, and conflict; Hirschhorn's *Managing in the New Team Environment* contains important links to psychoanalytic group theory; Bushe and Shani's *Parallel Learning Structures* presents a seminal theory of large-scale organization change based on the institution of parallel systems as change agents; and Hitchin and Ross's *The Strategic Management Process* looks at the connection between strategic planning theory and practice and implementation through OD interventions.

As editors, we have not dictated these connections, nor have we asked authors to work on higher-order concepts and theories. It is just happening, and it is a welcome turn of events. Perhaps it is an indication that OD may be reaching a period of consolidation and integration. We hope that we can contribute to this trend with future volumes.

New York, New York   Richard H. Beckhard
Cambridge, Massachusetts   Edgar H. Schein

OTHER TITLES IN THE ORGANIZATION DEVELOPMENT SERIES

Parallel Learning Structures: Creating Innovations in Bureaucracies
Gervase R. Bushe and A. B. Shani, 1991 (52427)

The Strategic Management Process: Integrating the OD Perspective
David Hitchin and Walter Ross, 1991 (52429)

The Conflict-Positive Organization: Stimulate Diversity and Create Unity
Dean Tjosvold, 1991 (51485)

Change by Design
Robert R. Blake, Jane Srygley Mouton, and Anne Adams McCanse, 1989 (50748)

Organization Development in Health Care
R. Wayne Ross, 1989 (18364)

Self-Designing Organizations: Learning How to Create High Performance
Susan Albers Mohrman and Thomas G. Cummings, 1989 (14603)

Power and Organization Development: Mobilizing Power to Implement Change
Larry E. Greiner and Virginia E. Schein, 1988 (12185)

Designing Organizations for High Performance
David P. Hanna, 1988 (12693)

Process Consultation, Volume I: Its Role in Organization Development, Second Edition
Edgar H. Schein, 1988 (06736)

Organizational Transitions: Managing Complex Change, Second Edition
Richard Beckhard and Reuben T. Harris, 1987 (10887)

Organization Development: A Normative View
W. Warner Burke, 1987 (10697)

Team Building: Issues and Alternatives, Second Edition
William G. Dyer, 1987 (18037)

The Technology Connection: Strategy and Change in the Information Age
Marc S. Gerstein, 1987 (12188)

Stream Analysis: A Powerful Way to Diagnose and Manage Organizational Change
Jerry I. Porras, 1987 (05693)

Process Consultation, Volume II: Lessons for Managers and Consultants
Edgar H. Schein, 1987 (06744)

Managing Conflict: Interpersonal Dialogue and Third-Party Roles, Second Edition
Richard E. Walton, 1987 (08859)

Pay and Organization Development
Edward E. Lawler, 1981 (03990)

Work Redesign
J. Richard Hackman and Greg R. Oldham, 1980 (02779)

PREFACE

All forms of social organization have two simultaneous needs that are often at odds with each other: freedom and order. Freedom springs from intuition and leads to innovation. Order stems from intelligence and provides efficiency. Both are essential, but are they compatible with each other? Within organizations, these requirements are translated into structural terms with which we are rather familiar. Freedom is translated as the specialized interests of different parts of an organization; the optimal goal of decentralization. Order is represented as the regulation and integration of all elements in harmonious and common action; the optimal goal of centralization. The problem with the centralization–decentralization debate, however, was that the more you realized the benefits of the one, the less you got the benefits of the other. The dilemma of organization was the dilemma of an either–or world, of being either a boss or a subordinate. The promise of a release from the dilemma, of the flexibility of both centralization and decentralization, specialization and integration. The price, however, is real—it is greater management and organizational complexity.

New York   S. M. D.
Boston   P. R. L.
September 1977

ACKNOWLEDGMENTS

The authors have received several kinds of generous help in the preparation of this book which we very much wish to acknowledge. Without being capable of identifying all of our many intellectual debts, we do want to especially note that Jay Galbraith has contributed to our thinking about information processing, and Jay Lorsch to our thinking about simultaneous decision making.

A considerable number of managers have of course contributed their insights and observations to the preparation of this work. Without being able to acknowledge all of these sources of help we do want to express our special thanks for the help of Gosta Almqvist, Jack Carlson, Paul Collins, Sheldon Davis, Keith Glegg, William Goggin, John Rudy, George Vojta, and Thomas Theobold.

For their assistance in the preparation of the cases used in the book, we thank Martin Charns, John Gabarro, Andre Reudi, Paul Thompson and Warren Wilhelm.

Finally, we appreciate the patience and care that was taken by Sarah Collins and Alice Mandel in helping to prepare the manuscript.

CONTENTS

1
THE MATRIX ORGANIZATION

Matrix management and organization are spreading in the United States and to a lesser extent in other countries. The list of well-known firms that are involved is becoming long and impressive. Take, for example, a company that has annual sales of $14 billion and employs around 400,000 people in scores of diverse businesses—General Electric. For decades, despite the diversity of businesses, GE used one basic structure throughout its organization:

GENERAL
MANAGER

FIVE FUNCTIONAL MANAGERS

In recent years some of its groups, divisions, and departments have adopted the matrix. The logic is that one must organize to meet the particular needs of each business and, if different businesses have different needs, then one organization design for all is bound to be inadequate. The matrix is appearing as a fundamental alternative. In a projection of GE's organization over the next ten years, its Organization Planning Bulletin (Sept. 1976) states,

> We've highlighted matrix organization. . . not because it's a bandwagon that we want you all to jump on, but rather that it's a complex, difficult, and sometimes frustrating form of organization to live with. It's also, however, a bellwether of things to come. But, when implemented well, it does offer much of the best of both worlds. And all of us are going to have to learn how to utilize organization to prepare managers to increasingly deal with high levels of complexity and ambiguity in situations where they have to get results from people and components *not* under their direct control. . . . . Successful experience in operating under a matrix constitutes better preparation for an individual to run a huge diversified institution like General Electric—where so many complex, conflicting interests must be balanced—than the product and functional modes which have been our hallmark over the past twenty years.

This expresses our own sentiments very well. A few years ago when we would ask a group of diverse executives if their organizations used a matrix only a scattering of hands would appear. Such a question today usually draws a response from well over half of the class. On further questioning it also becomes clear that not all of those responding have a very clear or consistent idea of what they mean by matrix, and a fair number will have reacted very negatively to their initial experience with a matrix. So the matrix approach is being tried in more and more organizations, not because of, but in spite of, limited understanding and mixed reactions.

These two facts, the spread of matrix and the confusion about it, explains why the authors, as teachers and researchers of organizational behavior, have for a number of years had a keen interest in studying matrix organizations. Our interest has culminated in this book. There are many important questions about matrix that we believe need answering even though our present answers must be tentative. The statements in this book about matrix are worded in a declarative form as a convenience of expression, but all should be viewed as our best current hypotheses subject to further testing.

WHAT IS A MATRIX?

The term matrix grew up in the United States aerospace industry. It probably seemed like a fitting term for mathematically trained engineers in that industry to apply to the gridlike structure that was evolving from its project management origins during the 1950s. Regardless of its origin, it has now become the accepted term in both business and academic circles. But how can we best define it? We believe that the most useful definition is based on the feature of a matrix organization that most clearly distinguishes it from conventional organizations. That is its abandonment of the age-old precept of "1 man—1 boss" or a single chain of command in favor of a "2-boss" or multiple command system. So we define matrix as any organization that employs a *multiple command system* that includes not only a multiple command structure but also related support mechanisms and an associated organizational culture and behavior pattern. The full meaning of this definition will, of course, develop throughout this book, but we must emphasize here that moving to a matrix organization, as we define it, is a truly significant organizational step. It is distinctly not just another minor management technique or a passing fad. For those business organizations who need a matrix and use it fully, it represents a sharp break with earlier business organization forms. To borrow a biological analogy, matrix represents a new species of business organization, not merely a variant of an existing type. But, is it truly new when we consider the full range of human institutions? We need to put the matrix into perspective.

ORGANIZATIONAL CHOICE

Business organizations have evolved since the start of the industrial revolution as "one boss" unitary command structures. This is so close to being a universal pattern that managers have been unaware until recently that any choice existed in this matter. Just the mention of the 2-boss idea made many managers distinctly uncomfortable. It seems to them like a violation of natural law, like rewriting the tablets from Mt. Sinai. But in even the most traditional businesses, the 1-boss rule is often strained. We have often heard such statements as:

Somewhere we've got an organization chart that tells you who my boss is, and there's a dotted line going the other way, but I really couldn't say which is which. They're both my bosses, and I doubt if either of them could tell you which is the real one.

Officially, I'm supposed to report to corporate staff, but you know for all practical purposes my real boss is our plant manager.

They say it isn't so, but I know that it is. Sometimes I feel that I've got three or four bosses.

More and more managers feel as if they report to 2-bosses regardless of the unitary organization chart. Many who do, and most who don't, however, see something messy and troublesome in the notion; as the quotes above suggest, it smacks of indecision and confusion. So many managers find themselves practicing in reality what they reject in theory. Some of these managers are going further to recognize that there is more than one theory, more than one model; that there is an organizational choice. Our ideas about how to organize in business derive from other institutions such as the military, religion, government, and the family. Although the tasks performed in each vary widely, all of them create organizations that distribute power to carry out their purpose. The distribution of power may begin through one or through many sources, and this distinction creates two basic organization designs that we will simplify by calling them the 1-boss model and the 2-boss model.

1-Boss Models

The *military,* the *church,* and the *monarchy* are all institutions that ɔelieve in and maintain pyramid-like structures whose plumbline is the unity of command. It took nothing less than the convulsions of the Protestant Reformation to create an alternative to the singular hierarchy of the Catholic Church. Despite the pluralism in religious thought and organization, however, business still adopted the unitary belief: thou shalt have but one boss above thee. To create the separation of powers inherent in the English and especially the American forms of government, we experienced the beheading of a king and revolt of the colonies. And these changes took decades, if not centuries, to unfold. Why, then, should business institutions be any more amenable to fundamental change in the distribution of power within their organizations than religions and governments have been to the same process within theirs?

Indeed, military organization provided the most important model for early business structures, and the scalar principles (power increases as you ascend in a hierarchy, superiors coodinate their subordinates' efforts) and notions of line and staff are still very much alive in both military and industrial complexes. Yet, the ironic fact is that it is at the interface between military and industrial organization, in the aerospace industry, that we today most frequently find matrix organizations. The military, in its need to have a single liaison with any one project in industry, are responsible for prompting a second managerial line and hence a pluralist model of managing.

2-Boss Models

While the 1-boss model accepts the greater authority of those higher in the hierarchy as a given, the 2-boss or multiple-boss model does not relinquish the subordinate's autonomy quite so easily. Hierarchy of power and status is not denied, but it is made plural. The best known examples of this model are found in our *families* and in our *government*.

Managers who feel or would feel uncomfortable in a 2-boss relationship might do well to remember that we each had both a mother and a father. As children, we were responsible to both of them, both had authority over us and, oedipal problems notwithstanding, the arrangement was basically comfortable. Parent–child conflicts are rarely due to the *existence* of a second parent; in fact, one parent often eases difficulties the child experiences with the other parent. As adults, most people are still able to negotiate their relationships with both parents and not feel uncomfortable about the duality.

Despite the significant differences between having two parents and having two bosses, nevertheless there are common elements in both situations: (1) parents and bosses both occupy superior positions in their hierarchies and (2) both must share their authority and status over a common subordinate(s). From the child's perspective, when resistance develops it is likely to be over the first element—the drive for independence and away from subordinate status. For the subordinate adult manager, however, resistance is caused by the second element—multiple masters.

To move to another model, relations between nation states have long been handled as a balance of power or what economists call an oligopoly. In an oligopoly, members agree to play by rules which will limit the distribution of wealth to only a few, giving up the goal of gaining complete control in exchange for a guarantee that they will never be excluded from having significant control. Oligopoly among firms is an economic counterpart to a balance of power between nations. Nation states have achieved long periods of peace by such rules, as firms have achieved long periods of prosperity by the same principles applied to economic activities. The balance of power concept, however, has had greater acceptance between nations than within nations. However, the American form of government, among others, is an exception.

The separation of legislative, executive, and judicial powers in the government is a 3-boss model. It is intended as a safeguard of individuals' liberty in the face of a government's need to maintain order. The tension between these simultaneous needs of independence and authority can be translated into terms that affect business organizations, as well. Business organizations need to divide labor into specialized tasks *and* to coordinate these tasks for the good of the total corporation. The freedom that comes from decentralized organization must be balanced with the integration and control of centralized forms, much as in the forms of government.

Americans in business are traditionally suspicious of the long arm of government and generally regard its methods of organization as inefficient bureaucratic nightmares. That government's reliance on a balance of power model could have useful counterparts in the organization of a business enterprise usually provokes the comment, "What profit did the government ever make?" The point is well taken, but so is the rebuttal: business has used the military for its model, and what profit has the military made? Different institutions can and do use the same model for different purposes.

The important point is that the spectrum of human institutions offers us choice in regard to organizational models. Business firms are now opening up this question and a considerable number are moving from the 1-boss to the 2-boss or multiple boss model. The realization of choice, however, only serves to raise a number of important questions.

QUESTIONS ABOUT THE MATRIX[1]

Why Is Matrix Used?

This question will be addressed in detail in the next chapter but an example of why an organization adopts a matrix can help set the stage. Let us assume we are managing a medium-sized ($50 M sales) growing firm in the specialized electronic instrument field. To be successful in this field one must stay at or close to the edge of the rapidly advancing technology and also stay in close touch with the specialized and changing needs of approximately 100 companies who are potential customers. The firm grew up with a conventional functional organization with each of the major functions of manufacturing, sales, and research and development headed up by one of the original founders. Now all of this original group have retired except the present board chairman. The present CEO advanced to the presidency through the ranks because of his conspicuous executive ability and now has become the center of all-important decision making in the firm. And that is the rub. Until recently, because of the rich informal ties forged during the founding years, the management system operated smoothly and quickly around the central figure of the CEO. But now the proliferation of product lines and the problems of size have erected a pile-up of issues awaiting the attention of the overworked CEO. Furthermore, there are increasing signs of bickering and fault finding between the functional departments, much of which are coming up around unexpected delays in the introduction of the new products that are essential for keeping a competitive advantage. The senior managers are aware that their most successful new product introduction of the past three years was a direct result of an informal cross-functional team that grew up around a middle-level sales manager who seemed to have the right kind of leadership touch. It is in this context that the top management team has begun to explore the matrix idea.

Based on our experience with matrix, we would say, "Right on." This is *one* of the kinds of situations that are a natural for evolving toward a matrix. In Chapter 2 we will spell out in more analytical terms why this kind of situation calls for a matrix. We will emphasize, however, that a matrix is an exceedingly complex organizational form that is not for everybody. To put it bluntly, if you do not really need it, leave it alone. There are easier ways to manage organizations. Or, as it says on the drug bottle, take only as directed. We will need to understand when *not* to use a matrix as well as when to use it.

How Does One Establish a Matrix?

In Chapter 3 we will examine the more common ways that firms have evolved to a full-blown matrix organization. Not surprisingly, the path followed seems to be largely determined by one's starting place in terms of the type of industry, size, etc. Tracing these various paths toward a matrix has more than a historic value since it emphasizes that a matrix, to be successful, needs to be "grown" through successive phases. It is hard to find organizations that have successfully jumped in one dramatic structural leap from a conventional single-chain-of-command structure to a complete matrix form. A lot of learning is needed before an organization can achieve a fully functioning matrix, and learning takes time and effort.

How Does a Successful Matrix Work?

This question will need to be addressed at several levels if we are to answer it with any success. What varieties of matrix have developed successfully? What are the distinguishing processes that are found in a matrix? How is the working life of key managers in a matrix different from that in a more conventional organization? What is it like to work for two bosses? What is it like to share a group of subordinates with another manager? These are some of the questions that are taken up in depth in Chapter 4.

How Does a Matrix Impact on Individuals?

This question brings the matrix close to home. Which kinds of people can work successfully in a matrix and which kinds cannot? What demands and pressures does a matrix place on people? This is not simply an academic question: in one firm a matrix that was succeeding in economic terms was dropped because the key professional staff felt it was just too stressful. Can such stresses be handled? Are there ways to educate people for the kinds of behavior called for in a matrix? These questions are answered in the affirmative in Chapter 5.

What Are the Hazards of a Matrix?

As is true of any major innovation such as the matrix, there are numerous ways to foul things up. While it would be exceedingly rash for anyone to claim to have catalogued all of these potential pitfalls, Chapter 6 focuses on the more common ones that we have had a chance to observe. The purpose of our review of matrix pathologies is, of course, not to scare off those who would otherwise benefit from matrix but clearly to provide the cautions that, it is hoped, will make prevention possible or at least provide a basis for a cure.

Who Uses a Matrix?

The matrix is such a recent and growing development in organizations that it is not easy to be precise about the scope of its present use. There is also a great deal of unevenness in the extent to which any given organization utilizes the matrix. It is clear, however, that examples of it can be found in a wide range of industrial and organizational types. We are aware of its use in the following areas:

Manufacturing Organizations

    Aerospace
    Chemicals
    Electronics
    Heavy equipment
    Industrial products
    Pharmaceuticals

Service Organizations

    Banking
    Brokerage
    Construction
    Insurance
    Retailing

Professional Organizations

    Accounting
    Advertising
    Consulting
    Law

Nonprofit Organizations

    City, state, and federal agencies
    Hospitals
    United Nations
    Universities

This is undoubtedly only a partial list but it does serve to demonstrate the breadth of application of the matrix idea. The spread of the matrix has happened with a minumum of public attention. This quiet revolution continues and it is not easy to predict when it will end. In Chapter 7 are presented some sketches of how the matrix idea has been adopted in a wide range of organizational settings.

How Is a Matrix Relevant to Multinationals?

Many managers believe that the matrix is uniquely relevant to multinational firms. We see its utility in bridging domestic–international differences and in maintaining both product and market area perspectives. We hasten to add that it is especially difficult to achieve a smoothly functioning matrix in a global arena. After all, distance alone is a barrier to the kind of interpersonal give and take that characterizes matrix decision making. These and many other special issues arising in the multinational matrix are discussed in Chapter 8. Furthermore, not all cultures are as amenable to matrix as in the United States culture, and in Chapter 3, under Cultural Patterns, we examine England, France, Germany, Japan, China, and Russia to find out why.

Where Is the Matrix Headed?

Many organizations have already adopted the matrix model and we believe that many more will in the future. But, even for these firms we do believe that the matrix is not necessarily a final organization form. The future of the matrix is difficult to predict. In spite of the difficulty we are exercising the license of authors to do some reasoned speculating about that future in our final chapter. The final verdict on the utility of the matrix approach is far from rendered, but commitments need to be made in the present. We must, therefore, make our educated guesses about the longer term potential of the matrix, and that we have done.

NOTE
1. Case studies appear at the ends of Chapters 2–8 as examples of discussions in the text. The general reader may wish to skip these case studies, read only a few, or read all of them. They are intended as complements to the text in each chapter.

2
THE MATRIX ORGANIZATION—WHO NEEDS IT?

In Chapter 1 we defined the use of multiple command as the essence of matrix organizations. But who really needs multiple command structures and management? Why not stay with the simple single chain of command? At this point we need to remind ourselves that every organization, based on matrix or not, is set up as a way of inducing the desired work behavior on the part of its members. This chapter will address the question of what actual organizational behavior one is trying to induce by using the matrix model as against the more conventional types. Even more importantly we will address the "why" question—under what environmental conditions would one want to induce these "matrix" behaviors? When is the matrix a sensible and practical way to bridge between the specific requirements for healthy survival that are thrust upon an organization and the actual work activities of organizational members? When does such a model serve to channel people's energies into needed tasks.

Our study of these questions has led us to the conclusion that the matrix is the preferred structural choice when three basic conditions exist simultaneously. This chapter examines these three conditions and their connection to matrix structures, systems, and behavior.

CONDITION 1: OUTSIDE PRESSURE FOR DUAL FOCUS

One of the principal reasons people form organizations is to focus attention and energy on a selected goal. Organizations serve as a lens that catches the sun's rays and bends them into a spot of focused energy. This is the source of the power of organizations and their leaders. It is why organizations can undertake tasks that are "too big" for a single individual or a simple small group. The initial way organizations focus human energy is to group people physically into different organizational units each with a defined boundary and a common boss. These groups are formed around a theme or symbol that identifies their purpose to the rest of the organization. Group members develop their own distinctive way of thinking, working, and relating to each other. They share a common task to be performed as their contribution to the work of the entire organization.

At a simple level, tasks can be "too big" for an individual for two basic reasons. The first is simply because an individual cannot be in two places at the same time. The second is because our mental capacity is finite—one individual cannot be expert and skilled in everything. (If this book had been written a century ago we would have had to add a third basic reason—that human physical strength is finite. But the spread of powered machinery has for all practical purposes removed this constraint.) When organizations initially form to get around the first constraint, they very naturally tend to set up their organizational units in terms of different physical locations. This tendency is especially apparent in transport and communication companies such as railroads, postal services, telegraph, and telephone. The same principle is at work when we assign a group to look after a defined set of customers. We expect to have the resulting organizational units bear names that identify both the location or set of customers and the service or product provided. We are familiar with the Milwaukee Airport, the New England Telephone Company, Saks Fifth Avenue, etc. This type of organizational grouping tends to focus attention and energy on performing the entire task or service for a given area or set of customers.

When organizations form to get around the second constraint (mental limits) they tend to group people initially around technical specialties so that the group members can enrich and reinforce one another's technical proficiencies. In conventional language this is known as a functional organization that identifies its primary groups with words such as manufacturing, sales, engineering, purchasing, finance, personnel, etc. These labels serve to orient each group to one technical specialty and focus energy accordingly.

Each of these ways of establishing a division of labor is widely used and each division has performed successfully over many years. But which is best, organizing around functions, areas, products, or services? It quite clearly depends on which constraint is more critical. If the geographical coverage is absolutely essential to the existence of an organization such as the telephone service, then it is wise to organize first of all into geographic units that focus on providing a complete service responsive to the special needs of customers in its area, even at the expense of the potential depth of expertise that could be achieved by grouping by technologies. Likewise, when technical expertise is critical to an organization's existence, then initially using functional (technology) groupings is sensible and perhaps essential. In such an arrangement, organizational power will center in these functional groups and the state of technical proficiency can be expected to advance, even at the expense of providing services and products tailored to the special needs of a particular locality or set of customers.

So each mode of organizing has its special strength and its corresponding weakness. But what if both types of constraints are truly critical and equally compelling? What if focusing attention on both is essential to survival? This is, we suggest, the first of the three basic conditions that call out for some form of a matrix design.

It is no accident that matrix first came into widespread use in the aerospace industry. To survive and prosper in the aerospace industry, any firm needs to focus intensive attention *both* on complex technical issues and on the unique project requirements of the customer. These companies can not afford to give a second-level status to either the functional groupings around technical specialties or to the project groupings around unique customer needs. They need to create a balance of power between project-oriented managers and the managers of the engineering and scientific specialists. Neither can be allowed to, arbitrarily, overrule the other. Both orientations need to be brought to bear in a simultaneous fashion on a host of trade-off decisions involving schedules, costs, and product quality. The needed behavior is epitomized by a picture of two middle managers with equal power, but very different orientations and goals, sitting down to debate and argue over each and every point in their search for the answers that would optimize decisions for both technical excellence and unique customer requirements. The dual command structure of a matrix serves to induce this kind of simultaneous decision-making behavior. It was to induce this kind of behavior that matrix was developed. The case study of Printer that appears at the end of this chapter presents an example of these outside pressures for maintaining a simultaneous dual focus in an organization that is only beginning to move into a matrix.

CONDITION 2: PRESSURES FOR HIGH INFORMATION-PROCESSING CAPACITY

The second condition that generates pressure to adopt a matrix is the requirement for high information-processing capacity among organizational members. Once any organization is formed to do work "too big" for individuals, it must pay the basic price of organizing—it must establish and maintain a network of communication channels among members. When only one person is doing a job a single nervous system is used to keep the right and left hands coordinated. When many people are involved in a task, the extra "overhead" cost of coordinating the messages sent back and forth between people must be borne. Since communication uses resources, organization planners try to arrange clusters of people and channels between them to minimize the cost of required communications. The hierarchical pyramid of a conventional organization, depicted with its boxes and lines, represents an attempt to conserve resources by channeling communications through selected managers. Such a communication hierarchy can be supplemented by rules, job descriptions, standard procedures, schedules and budgets which, in addition to personal instructions from the boss, indicate to members what behavior they are expected to engage in that will fit in a coordinated way with the work of others. These coordination arrangements work fine, if they do not get overloaded with information. But under certain conditions they do get very overloaded.

The symptoms of such overloading are familiar to managers. The issues urgently awaiting managerial action pile up. The queue to see the boss gets long. Schedules and budgets start slipping but nothing gets done about it. Bureaucracy sets in. There are too many rules, it is felt, but more probably it is that the channels are not organized properly. The communications process bogs down. In effect, the right hand loses track of what the left hand is doing.

Why do smart, hardworking, well-intentioned managers some-times get themselves into this situation? Sometimes, better schedules, better budgets, bigger computers, better rules, and clearer job descriptions can cure the problem. As often as not, however, they are an inadequate cure. There are still too many real issues that have to get resolved and not enough hours in the day to resolve them. Under such conditions only a fundamental redesign of the organization can relieve the information overload. What conditions tend to generate an overwhelming need for information processing and complex problem solving? Only a special combination of circumstances can lead to a very high information-processing requirement.

First, the kinds of demands placed on the organization have to be changing and relatively unpredictable. If the demands are stable, and therefore reasonably predictable, there is not much important new information for the organization to cope with. Plans can be made in advance and the assumptions about future events that the plans were made for will prove valid. There will seldom be a need for quick replanning. Things can go according to schedule. But the future frequently holds major surprises such as sudden changes in market demand, competitive moves, technological advances, ecological restrictions and other governmental regulations, currency and stock-market fluctuations, and the appearance of protest groups. When these surprises occur, plans do not hold up, and large amounts of new information must be assimilated and responded to in a coherent way. *Uncertainty* in the external environment calls for an enriched information-processing capacity within the organization.

Second, even these uncertainties would not be unmanageable if one's organization was doing a simple job such as making and marketing a single product in a single area, or providing a single service to a single customer. It should be remembered that, before the matrix, the last major change in organization design occurred in the 1920s and 1930s when businesses diversified their activities in both product and market terms, leading to the shift from centralized functionally departmentalized organizations to decentralized ones based on a product division design. While the increased complexity of tasks led to a major adaptation in terms of management and organization, both the centralized and the decentralized models maintained the traditional singular chain of command. Information and communication were organized either along functional lines, or by product category or market unit. Simultaneous diversification of both products/services and markets, however, increased the *complexity* of an organization's tasks severalfold. When this complexity is "added" to environmental uncertainty, the result is a major multiplication of information processing requirements.

Finally, the question arises concerning how many individuals and groups must be involved in order to make a reasoned response to new events. The more *interdependence* there is among people on any one issue, the greater the information-processing load. If people can accomplish their tasks, no matter how complex and uncertain, by themselves, then they will not have to share information with others and the information-processing load will not be great. If, on the other hand, their tasks are highly interrelated, the opposite is bound to be the case.

So all three generators of the information load—uncertainty, complexity, and interdependence—have to be examined. If all three are high, conventional ways of handling the load tend to break down. If such a compound piling up of information-processing requirements were a rare, once-in-a-hundred occurrence, we could afford to ignore it. But it appears that this set of circumstances is to be the fate of more and more organizations, even in industries that we could label as stable and mature in the recent past. When organizations have to come to terms with heavy information-processing loads, they have to open up and legitimate a more complex communication and decision network.

If the problem were only one of keeping more people informed of events, the response could be handled through increasing the flow of reports, briefings, and informal communications. But, of course, the tough part is weighing the significance of the new information and making decisions that commit the organization to a response that will prove to be wise over time. To accomplish this, more people simply must be in a position to think and act as general managers—more people who seek out and pull together the relevant information and opinion, who weigh alternatives, who make commitments in the best interest of the whole and who stand ready to be judged by the eventual results. This is the kind of behavior that is called for when managers handle large amounts of complex information for the organization. This behavior on the part of more people is the untimate cure for information overload.

The matrix design, properly applied, tends to develop more people who think and act in a general management mode. By inducing this kind of action, the matrix increases an organization's information-processing capacity.

The case study at the end of this chapter provides an example of overload and the case studies in the next several chapters describe the information-processing capacity of matrix at work.

CONDITION 3: PRESSURES FOR SHARED RESOURCES

The third and final condition we see as an indication to adopt a matrix is so obvious that its importance is easy to ignore. It is the condition of the organization's being under considerable pressure to achieve economies of scale in human terms and high performance in terms of both costs and benefits by fully utilizing scarce human resources and by meeting high-quality standards.

While all organizations probably feel some degree of pressure for high performance, the amount of such pressure varies a good deal. If an organization has a truly dominant position in a given market, it may not feel much pressure to avoid redundancies in the use of human resources or to attain high-quality outputs. It may feel comfortable increasing its costs by hiring extra specialists to fill out the needs of multiple product divisions, even though these specialists wind up spending much of their time unproductively. On the other hand, it could hire one jack-of-all-trades to perform the jobs of several specialists with a corresponding loss of quality. But when performance pressures are real and strong, the need arises to fully utilize expensive and highly specialized talents. The size of an organization may enable it to acquire skilled human resources in large numbers, but nevertheless there is bound to be an upper limit. Whatever that limit is, when it is reached, pressures will develop to share existing human resources. These internal pressures will be greater when there are external forces pressing to achieve economies of scale. These resources will need to be redeployed in a flexible manner so that people can work on more than one task at a time or at least be readily available for assignment from one task to the next.

A similar argument holds for the sharing of expensive capital resources and physical facilities. For example, several product divisions may need access to a fleet of different test aircraft (helicopters, propeller-driven and jet aircraft) to evaluate their products but no divisions can afford to maintain them full time. High performance will result from high utilization of such facilities through effective sharing and redeployment of them among specialist groups.

Organizations with conventional designs tend to develop resistances to the rapid redeployment of specialists across organizational lines. Structures are traditionally thought of as stolid and static. They do not change very often, and when they do it is in a discrete quantum jump to another static state. Since environmental and strategic changes tend to evolve in a continual process, the organization is often catching up with already changed circumstances. And each change is experienced by its members as a wrenching of established patterns of behavior and the need to learn new ones. Also, the more rapidly the environment and the firm's strategy change, the shorter will be the duration for which a given structure is appropriate. In such circumstances, it is helpful to think of optimal structural change as being frequent and in small doses, rather than infrequent major shake-ups. Structure, then, becomes flexible, if not fluid; and people can become accustomed to a structure that is always changing but that rarely erupts and causes severe dislocations. The matrix design helps induce the kind of behavior that views rapid redeployment and the shared use of scarce human resources as basic.

REVIEW: THE THREE NECESSARY AND SUFFICIENT CONDITIONS

Based on our observations of why managers adopt a matrix, we initially stated that all three conditions discussed above need to be present simultaneously before a matrix is indicated. Why is not the presence of one or two of these conditions an adequate reason? Our analysis runs as follows: learning how to use a matrix is not easy. A matrix organization is not simply a matter of understanding and creating a formal design. For us:

$$\text{Matrix Organization} = \text{Matrix Structure} + \text{Matrix Systems} + \text{Matrix Culture} + \text{Matrix Behavior}.$$

The structure involves the dual chains of command that we have spoken about. The system must also operate along two dimensions simultaneously: planning, controlling, appraising and rewarding, etc., along both functional and product lines at the same time. Moreover, every organization has a culture of its own and, as we shall discuss more fully in Chapter 3, for the matrix to succeed the ethos or spirit of the organization must be consonant with the new form. Finally, people's behavior, especially those with two bosses and those who share subordinates, must reflect an understanding, and an ability to work within such overlapping boundaries.

The change to a matrix cannot be accomplished by issuing a new organization chart. People are brought up, by and large, to think in terms of "one person, one boss" and such habits of mind are not easily changed. People must learn to work comfortably and effectively in a different way of managing and organizing. Our experience suggests that successful passage through the early evolution of a matrix, until it is firmly established in its mature form, is a process that will likely take two or three years. That is not a long time in the evolution and life span of an organization, but to those involved in the change, the period of transition can be quite difficult. Furthermore, the limited evidence suggests that going to a matrix will initially add to managerial overhead. So one is ill-advised to view the matrix as one of many exciting new managerial tools and techniques; something that can be tried and discarded if it doesn't seem to be succeeding. The move to a matrix should be a serious decision, made by the top level of management, signalling a major commitment, and thoroughly implemented through many layers of the organization. It is too difficult to undertake superficially, too costly in human terms to attempt haphazardly, and too encompassing to experiment with unnecessarily.

The presence of only two of the three necessary conditions is not sufficient for us to recommend a matrix approach. There are simpler, partial methods for coping with the additional needs. Taking each partial set of conditions separately:

- It is clear that without performance pressures, the problems generated by conditions 1 and/or 2, the pressure for dual focus and for high information-processing requirements, could be handled without a matrix simply by accepting a lower performance level. One could either take whatever time was needed in a conventional organization design to process all the information up to the top level where a dual focus would be brought into play—at the cost of very slow performance—or one could split the organization into decentralized autonomous parts thereby increasing the information-processing capacity, but at the expense of high costs or low quality.

- The presence of only conditions 1 and/or 3, pressures from two critical sectors and pressures to achieve human economies of scale and peformance, could reasonably be handled by creating a small top-management team that represented the dual focus and that given the limited informational requirements could then generate high quality and timely decisions.

- Finally, if only conditions 2 and/or 3 were present—to perform uncertain, complex, and interdependent tasks with scarce human resources—the problem could be solved by placing the most critical focus on the chain of command or line roles while using subordinate or staff positions to represent the less critical focus.

As summarized in Table 2.1, there are three environmental conditions, each of which calls for organizational response, and all of which must be present simultaneously for an organization to appropriately adopt and adapt the matrix.

Table 2.1   Necessary and Sufficient Conditions

	Environmental pressure	*Behavioral linkage*
Condition 1	Two or more critical sectors; functions, products, services, markets, areas	Balance of power, dual command, simultaneous decision making
Condition 2	Performance of uncertain, complex and interdependent tasks	Enriched information-processing capacity
Condition 3	Economies of scale	Shared and flexible use of scarce human resources

In summarizing the three conditions which call for a matrix, we must point out a very real hazard. It is easy for managers from all kinds of organizations to read over the three conditions and readily nod their heads in a quick agreement that all three conditions are present in their situation. Such as initial reaction is understandable. After all, most managers feel pressured from multiple sides and swamped

with information. If they felt otherwise, they would have trouble justi-
fying their salaries. But a facile reaction is dangerous. Organizations
do, in fact, vary in the extent to which they experience these pressures.
Until these three conditions are overwhelmingly present, in a literal
sense, the matrix will almost certainly be an unnecessary complexity.
Caution should be exercised in judging the presence or absence of
these conditions and one is well advised to err on the conservative side.
Clearly, only a limited, even if growing, number of organizations
really need a fully evolved matrix.

While the logic of the three conditions is clear, it may not be clear
just exactly how the matrix tends to induce the complex behavior that
can simultaneously meet all three pressures. Much of the remainder of
the book addresses this question, but at least a partial answer is called
for now.

The threefold behavior we are trying to induce with the matrix is
(1) the focusing of undivided human effort on two (or more) essential
organizational tasks simultaneously, (2) the human processing of a
great deal of information and the commitment of the organization to
a balanced reasoned response (a general mangement response), and (3)
the rapid redeployment of human resources to various projects, products, services, clients, or markets. Figure 2.1 can help in clarifying
how the matrix induces these behaviors.

We see here a diamond-shaped organization rather than the conventional pyramid. The top of the diamond represents the same top
management symbolized by the top of the pyramid. The two arms of
the diamond symbolize the dual chain of command. In the typical case
the left arm would array the functional specialist groups or what could
be thought of as the resource or input side of the organization. The
right arm arrays the various products, projects, markets, clients, services, or areas the organization is set up to provide. This is the output
or transaction side of the matrix. Depending on how many people
holding a specialist orientation, either resource or output, the organization needs, these groupings can develop several echelons in
response to the practical limits of the span of control of any line manager. At the foot of the matrix is the 2-boss manager. This manager is
responsible for the performance of a defined package of work. The
manager is given agreed-upon financial resources and performance
targets by superiors on the output side, and negotiated human and
equipment resources from the resource manager. The two streams,

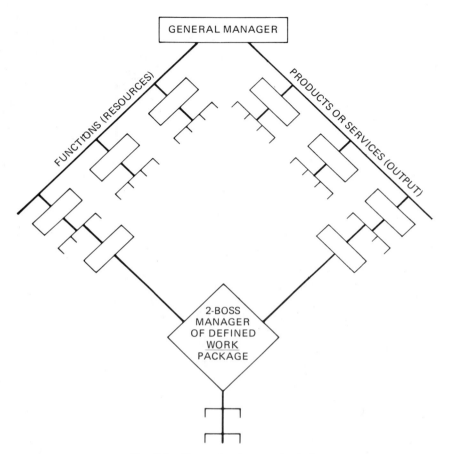

Fig. 2.1   Example of a matrix design.

taken together, constitute the work package. The manager is responsible for managing these resources to meet performance targets. To
perform, the manager must handle high volumes of information,
weigh alternatives, make commitments on behalf of the organization
as a whole, and be prepared to be judged by the results. This form of
organization induces the manager to think and behave like a general
manager.

This 2-boss manager is subject to dual demands of both the resource hierarchy and the transaction hierarchy, but can also draw upon their specialized help. The manager who cannot reconcile the dual demands is expected to convene a meeting with the two bosses and present the problem for the two of them to solve. In the resulting debate neither boss can, according to the authority structure, overrule the other. They must search for mutually agreeable and timely solution. While they can, if needed, have recourse to their respective two bosses, they cannot refer too many such disputes upward without reflecting on their own managerial capacity. The matrix induces many peer debates on key trade-off decisions. It tends to ensure that each of these decisions are made on their respective merits and not on the basis of any arbitrary form.

When the mix of tasks undertaken by the output or transaction side of the matrix shifts over time, the resources of the other side need to be redeployed. While it is never completely simple to move people from one task to another, the matrix can help this process along. The resource specialists can be engaged in a time-limited way to help any matrix product manager who needs their type of talent. This can be done over and over again without shifting the individual specialist from his or her "home base" resource group. People who are subject to redeployment in conventional organizations almost always develop a very understandable resistance to being uprooted and forced to join up with a set of strangers time after time. They cannot build a reputation for performance that carries them past the occasional mistake. They must work to create relations of trust and mutual respect again and again. While these problems are not totally avoided in a matrix, they can be greatly mitigated. Human resources can be redeployed with a minimum amount of the kind of associated human costs that also usually wind up as economic costs.

Finally, we must point out that even in a fully developed matrix organization, only a relatively small proportion of the total number of people in the organization will be directly in the matrix. Whereas a middle-level manager may have two bosses, those people reporting beneath that manager are likely to have only one boss. In an organization with 50,000 employees only 500–1500 may be in the matrix; and in one with 500 people, only 50 may be in the matrix. To keep in perspective the proportion of people that will be affected directly, it may be helpful to envision the diamond of the matrix perched on top of the traditional design of the pyramid. Drawn to scale, proportionate to the numbers of people involved in the matrix, the total organization chart might look like this:

The Printer case study which follows provides a concrete example of an organization that is experiencing the need for a matrix. The case study indicates how the matrix can make a major contribution to inducing the kinds of behavior described above. The matrix that is evolving at Printer can serve as the organizational link between the pressures of the environment and the three kinds of complex behaviors needed to respond to these pressures.

CASE STUDY: PRINTER INCORPORATED

In 1969, the management of Printer Incorporated were evaluating the recent change to "industry specialization" in the company's organization. The company's internal marketing and engineering departments, located at the Watertown, Wisconsin, headquarters, and several of its field sales regions had been reorganized by industry groupings of Printer's customers. Since the implementation of the organizational change, Printer's sales had increased substantially. One of Printer's executives commented:

> As long ago as 1964 we had a fairly clear notion of the need for industry specialization. Now, five years later, we have gone a long way down that road. By and large, it has been a good thing, but we are not sure if we are all the way down the road or only part way. Is the implementation of the change finished? Should we push further in the manufacturing area? Did we push far enough or too far in our field sales organization?

Founded in 1926 and based in the countryside of Wisconsin, near Milwaukee, Printer produced and internationally marketed specialized production printing machinery and supplies that were especially designed to print directly on such customer products as electronic tubes, apparel, and shoes.

Printer's customers were in several diverse industries, each of which had its own unique problems in printing directly onto its products a company or other type of product identification. Traditionally, a customer with a special printing problem came to Printer which met the customer's needs with a standard or specially designed machine and combination of printing supplies. Although in specific industries several companies competed with Printer, there were no other companies which sold printing products to customers in the wide range of industries that Printer served.

Printer's customers were in industries which were characterized by different product life cycles. For example, the shoe industry, for which Printer held a substantial part of the market for printing requirements, had few basic product changes and a very stable production process technology. From one year to the next it had generally the same printing requirements. The production process utilized little sophisticated equipment; and the operation of printing the company name, shoe size, and identification numbers could be handled as an additional batch operation in the process of shoe manufacturing. The electronics industry presented printing requirements at the other end of the spectrum. Rapid technological advances brought changing product sizes and characteristics, resulting in rapidly changing printing requirements. High-volume production utilizing sophisticated production equipment required high-speed printing equipment that could be integrated directly in the production process.

Thus, in serving the needs of these different types of customers, it was necessary not only to produce printing equipment and supplies that met the technical characteristics demanded by the customer's product characteristics but also to understand different customers' production process requirements and characteristics of their industry.

Meeting the technical requirements of a special printing problem required the selection of and often design of a combination of machine, printing element, and ink or foil. Technical considerations in making this choice included such factors as the speed of operation, surface characteristics of the item to be printed, drying time requirements dictated by the production process, ability of the product to withstand heat of a drying oven or the necessity of air drying, and the unique characteristics of the combination of ink, printing element,

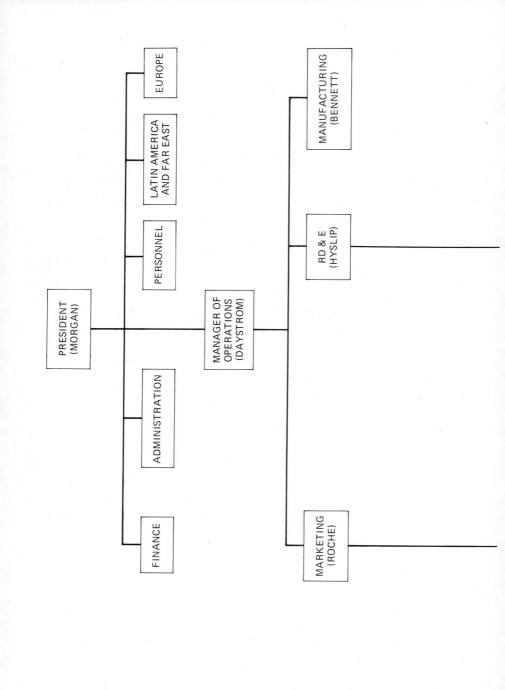

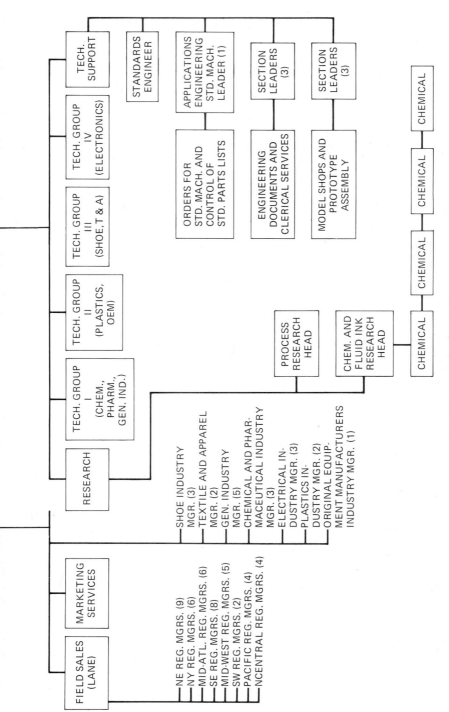

Fig. 2.2 Organization chart.

and machine. Improper selection of printing supplies and equipment resulted in problems such as smearing of the impression, inconsistency in printing quality, or gumming of the printing elements.

As shown in Fig. 2.2, the company was organized along the lines of marketing, engineering and research, and manufacturing. Of the 550 employees in the company, 170 were in the marketing organization. Several of Printer's managers explained this by stating that the company had a "marketing orientation." The Marketing Division was organized into a field sales organization of 85 people and an industry sales organization. The field sales organization, responsible for customer contacts for sales and service was organized geographically into eight regions. Each region had a regional office with regional manager and a small office staff and several regional field salesforces.

The industry sales organization, divided into seven industry groups, provided inside support for the field salesforce. The industry sales organization handled order processing and other sales record keeping, and worked with engineering and manufacturing to meet the needs of the salespersons' customers.

The Research, Development, and Engineering Division was organized into research, four technical groups, and some support groups. The research group concentrated on chemical research, and each of the four chemists worked primarily with one technical group. Each of the technical groups was responsible for engineering products for customers in one or two industries.

The Manufacturing Division was organized into machine centers (90 employees), equipment assembly (28 employees), printing elements manufacture (75 employees), ink formulation (28 employees), foil production (20 employees), and miscellaneous support groups (total of 50 employees). The ink formulation department prepared—often to customer order—over 10,000 different types of inks. The machining departments prepared from blueprints several different types of parts for stock, and the assembly department assembled these parts and those purchased from outside suppliers into equipment to fill customers' orders.

Prior to 1965 the Marketing Division was organized by region both internally and in the field, and the Engineering and Manufacturing Divisions were organized along functional lines. In 1965 the internal marketing department was reorganized along industry lines, but it returned to a regional organization in 1966. In 1967 it was once again reorganized along industry lines. Beginning in May 1968 the industry specialization concept was applied to the field sales organization, and shortly thereafter the technical division (RD&E) was reorganized. The "general industries" category applied to customers who were not in any of the other six industry groups. In the field sales organization, salespersons in most regions, beginning with the north central region, were given responsibilities for specific industries rather than for specific geographical areas in their regions. In a few instances in which there was a high concentration of one industry and a wide geographical dispersion of the others, the field sales organization still used a geographical breakdown of assignments. In those cases, the field salesperson had to contact more than one internal marketing group to meet customers' needs, whereas a purely industry-specialized field salesperson had one internal industry-specialized marketing group which served as an internal contact. At the time of the reorganization of the field sales organization, the functions of sales and service were specialized and assigned to separate personnel in several regions.

The internal industry groups and field salespersons were both paid straight salaries. The company had previously paid its field salespersons salary plus commission, but this was discontinued because management felt the additions and transfers of personnel and the differences in sales areas made the system inequitable. The company's information system, which reported sales of machines and supplies as compared to forecasted sales, was applied to both regions' and industries' performance.

Regional managers as a group were paid somewhat more than industry managers. Both the industry managers and regional managers were characterized by a wide range in age and length of service, with each group having at least one employee who had been with Printer less than four years and several who had been with the company over 20 years.

Several of Printer's management and employees commented on the implementation and functioning of the concept of industry specialization. James Roche, the marketing manager, commented:

> Our 1960 sales were 20 to 30 percent of today's. Most of this growth comes from growth in the industries Printer moved into. In 1911 it was shoes and textiles, in the '30s and '40s the electrical industry, in the '50s the pharmaceutical industry, and in the late '50s and '60s the explosion in electronics that pushed sales up. Top management is always looking for new markets. With 90 percent of the market in many industries, Printer is a leader as a broad-based marketing company.
>
> The printing compounds—wet ink or dry carrier-supported ink—are chemical, and machine design is mechanical with a small but increasing

electrical requirement. The common denominator in all the technical groups is an appreciation of printing processes—you have to know the whole spectrum.

Different industries have different development lead times. The electronics industry, for example, requires fast development and frequent calls on customers to keep up with their product developments. Customers in the electronics industry have a high technical ability themselves. At the other extreme the customer, say in the shoe industry, realized post facto that he needs to mark his products. You have to tell him how to do it. He wants you to do the whole thing.

I had the industry concept fixed in my mind when I came with the company. From previous experience with another company I feel it is the way to organize. Sales this year are up 20 percent from last year and are seven percent ahead of our goals, but we're having a hell of a time getting the products out. I would put a good measure of our success in 1969 due to our organizational change. The sales come from new products. We now have the technical people working and cooperating with marketing.

You can change things inside like a meat grinder and still survive, but you can't do it in the field. Many employees in the field didn't want to change to industry specialization, but they agreed to do it on a timetable starting first in the north central region. Most of the negative aspects of industry specialization are in the field. For example, it results in higher travel expenses. The field sales force has employees who have been with the company in their region for a long time. For them industry specialization means losing friends and contacts in those industries they would abandon in specializing. We took the stand of not specializing only where it is not practical—such as in the southeast region where the electronics industry is too scattered for specialization. We do have a good degree of specialization where it counts.

William Hyslip, manager of RD&E, commented:

This is a marketing-oriented company. In fact, it used to be that some of our engineers would take orders from nearly anyone in marketing without question. Industry specialization in both marketing and the technical group has created a more positive atmosphere. Now I think marketing feels they need the engineers. There are conflicts with field sales because they have a different orientation and they wonder what we are doing with our resources.

There are two types of work for engineering—customers' orders and new products. When the industry specialization came in, marketing was reorganized first, and they started dividing orders by industry. When engineering was reorganized, the industry technical groups got orders for their industries only, so they had to work with their corresponding industry group in marketing. Our best-selling machines have developed from a machine designed for a specific application and then extended. Shot in the dark development hasn't been successful. We don't want to build first and then try to sell, but rather design what the customer wants. We're doing this better because engineering and the internal marketing men are getting more contact with the customer. The engineering requirements are getting highly specialized by industry, and the engineers must understand the customer's application. Almost 100 percent of our design work is initiated by marketing.

Problems do come up in allocating engineering responsibility for machine design on machines used in more than one industry. Generally, we assign responsibility to one industry technical group. The engineers know pretty much what's going on in other technical groups, but the responsibility for this falls back on me. Maybe we should hold regularly scheduled meetings for this, but now we have irregularly scheduled ones and I see all the men almost each day.

The RD&E Division has about 60 people of whom one-third hold degrees, another third have taken a 2- to 3-year engineering course, and the remainder are "self-made." The industry specialization has allowed us to change the job content to give more of the men more of a "whole task." In contrast to a functional breakdown of work, we've tried at all levels to get development men to go with marketing men on customer visits.

George Healy, manager of the electronics technical group, commented:

About 75 percent of the engineering work in all industries uses some basic standard modules, but the applications are quite different and require different engineering skills. In organizing by industry, the men become stronger specialists through experience, but they also become less flexible. Another negative aspect about industry specialization is that when one group gets a surge in work, it may not have the manpower to cover it. One of the advantages of it, though, is that now people know where to go in engineering to get some action when there is a problem. Problems used to get lost in the department. It was too hard for the marketing man to know what a technical worker was doing, but now the marketing man gets more attention. We have a great relationship with our industry marketing group. Our priorities on work are based on Bill Williams' (industry marketing manager) knowledge of the industry and my estimates of how long it will take to do the work.

Tim James, manager of T&A and shoe technical group, commented:

Before industry specialization, our contact with sales depended on the project. There was one particular person in sales for each particular project. Now I get in right in the beginning on the specs. I'd write them if I could (marketing does); as it is I bias them pretty heavily. Marketing lets me have my way until a machine is pretty much designed, then they come up with things we didn't put in—like a Monday morning quarterback. We are in an experimental stage now.

People in my group are out in the field only one or two times a year, but the work load is too high to go out more. With the new industry groupings there is narrowed communications. We can get around the technical shortcomings by trading people among groups.

Ben Cabot, research manager, commented:

Having a chemically oriented man on the industry team facilitates its handling of specific problems and helps communications. They can solve problems earlier in the game. Sometimes you're unsure of whether a problem is with the ink or the machine. It requires a team effort attacking the problem from both ink and mechanical standpoints. The chemists have to work most closely with the industry technical groups, then with marketing, then manufacturing.

We need industry specialization. For example, in T&A one has to know dyes, dry-cleaning methods, history of the material, and technical terms of the industry. In electronics the terminology used to talk to the customer is quite different.

When a sticky problem is solved in one industry, it spills over into others. Bill Hyslip and I transfer information across industries. We sit on technical marketing meetings and switch manpower around.

Bill Williams, internal marketing manager—electronics industry, commented:

We had industry specialization a few years ago, but it was discontinued because someone favored geography over industry. Field sales was still organized geographically, and they had to talk to several people inside to get answers to all their problems. There were complaints from the field salesforce calling in to Watertown.

The job of sales correspondent (internal sales) hasn't changed much through the reorganizations. He is a salesman and must sell as much as the guy outside. He takes care of the customer and field salespeople that are in his area. He has to write letters, give quotes on machines and parts, handle customer problems, and take care of field salesmen's problems, such as when they need a new price book.

Industry specialization has given 1000 percent better communications between engineering, internal sales, and field sales. Giving the industry specialization to field sales makes it easier to communicate with the internal and field sales organizations. Marketing was specialized by industry first, and then engineering. When marketing was industry-oriented way back, I loved it. I kept pounding the table to get engineering changed to it.

Now the engineers can better understand what the customer wants. The salesman's job is to make sure the customer gets a product that fills his needs, but sometimes engineering doesn't understand the application. A lot of times their design doesn't use common sense, generally when they haven't been in the customer's plant. There used to be no connection between the jobs the engineers would get. Two engineers could get the same job for different customers and come up with different solutions. Out in the field we see things others have done. We see what doesn't work well and feed that information back to engineering—sometimes too late. My 25 years' experience can save a young engineer from designing something I know won't work.

One of the problems we have here is where a machine design works OK here at Watertown but after two to three weeks in production it falls apart. Before, we didn't know who to send. Now we take the man who engineered it. Generally a certain group in assembly does all the work for an industry, so we get together with them and engineering, and then we can send the engineer out with answer in hand.

Ed Clark, internal marketing manager-general industries, commented:

Industry specialization makes it easier to have someone to contact for specific problems, but in general industries, we need a project orientation. We have to cover too much area, and it is hard to determine what to do daily. I don't feel responsible for daily sales, although officially I am responsible—it is projects (new product introductions) that I feel responsible for.

Field salesmen have much more prestige in this company than do inside salespeople, and the regional managers have more prestige than the industry managers. They have direct access to the president. According to field sales, inside sales doesn't know how the real world works. But our role in making the company go is significant. We are the guys that are going to come out with the new products and spot problems with the old. We have a larger field to cover than the other industry groups, which makes it the hardest to handle but also gives the most flexibility. The company's opportunities to broaden its scope of markets occur in general industries.

Clark was asked if he had more trouble with field sales than engineering. He responded:

Oh hell, yes. I can talk to engineering. Communications are very bad with field sales. We just keep them off our backs in servicing their requests. I can't win any battles I fight with them. We are all on the defensive with respect to field sales. We rarely go out and try to help them, although some of that is happening now. I don't work very much with manufacturing—mostly on sales forecasting, and when they screw up. When they are late, we holler to our boss and he gives it to them. We don't understand their problems, and usually I don't talk to them on anything except when I'm in trouble. Engineering often relays the information to us from manufacturing.

Frank Leonard, north central regional sales manager, who had joined the company in the 1940s, commented on the implementation of industry specialization.

Back in the '50s we had industry specialization inside and we abandoned it for geographical organization. The fellows in the field were not specialized, and they had trouble talking to many people in the plant. They weren't big enough to specialize in the field. In 1964–65 we went back to industry specialization inside. Many people in the field were getting older, and we questioned what we should do to give them incentives and opportunities—how could we exploit their strengths in certain areas? In the north central region we found that 80 percent of our sales came from 20 percent of our customers. We looked at the interests of the nine salesmen and tried to fit them with particular industries. Most salesmen know best the line that they sell most—a guy could be good in five of 90 basic types of equipment. Some guys are more well rounded, and we use them for general industries.

The most valuable manpower asset we have is in the field sales force. We knew that these men—as is always true—would have a natural resistance to a major change such as this, which would interrupt their established responsiblities and routines. We knew we had a selling job to do with these men because, if they weren't enthusiastic, the plan was sure to fail, as they were the doers. The most effective thing we developed was an outline of the advantages to them, the company, and our customers. (See Fig. 2.3.) The reaction from the field salesmen varied from very negative to high enthusiasm, largely dependent upon the person's age, personal objectives, and capabilities. We lost no one. Only one man had to relocate, and he did.

Before the change, we were never able to get salesmen to schedule beyond one day. Because they had to do service, they couldn't plan. Now we have a separate service staff. The servicemen, who used to do less than one-half of the service calls, now do about 88 percent of them. On a survey two months after specialization, sales calls were up 22 percent.

Field sales has little contact with other parts of the organization. The regional sales office should be able to handle all of the standard line requests. The farther out you get from Watertown, the more of the customers' needs are handled by the regional office. By habit some of the customers call Watertown. The primary contact for the field salesman is the internal industry man, who handles all the needed inside contacts.

It used to be that a field salesman would find a good application that needed a new machine and he would get little response internally. Now there is more common interest between field salesmen and the internal people, and they get more interest and attention. I still get complaints from my sales personnel that their requests are not done well enough—there seems to be a variance in the quality of service provided by the different industry groups.

Joe Benzing, New York regional sales manager, who had been with the company for three years and had been regional sales manager for nearly two, commented:

ADVANTAGES TO CUSTOMER	DISADVANTAGES TO CUSTOMER
• Served by an industry-oriented, more knowledgeable salesforce • Better recognition and response by Printer to customers' needs • More attention from salesforce on total requirements — including consumable supplies	• Change of salesperson - initially • Possible slower response from sales personnel • Small customer receives fewer calls

ADVANTAGES TO PRINTER	DISADVANTAGES TO PRINTER

• Quicker market penetration for new products
• More efficient for new product introduction-fewer sales personnel involved at first
• Deeper penetration of existing markets
• More flexibility in a company or market growth situation
• More effective sales calls
• Better communication between:
      Sales and customer
      Sales and Watertown
• More field quoting
• Fewer field mistakes - credits
• Greater promotion of complete Printer package — more attention to consumable supplies
• Better administration and coordination of field activities by:
      Regional managers
      Industry and product managers
• More alert to industry changes (market intelligence)
• Easier and quicker to train salesforce
• Easier to assign employees within their capabilities
• Better selection and qualification of prospects
• Greater team effort among sales personnel
• Doing a better job in fewer areas
	• Slower response to customer requests
	• More sales travel time and cost-overlapping
	• More difficult to adjust work load of individual salesperson
	• Harder to deploy personnel geographically
	• Work force planning more critical — replacement or transfer
	• Possible higher turnover of sales personnel

ADVANTAGES TO SALESPERSON	DISADVANTAGES TO SALESPERSON
• Improved motivation - greater proficiency and satisfaction by being more knowledgeable • Improved technical training — easier to learn, and therefore become more effective, by concentrating on fewer products and industry areas • Improved job skills — opportunities to work in areas of greater interest and ability	• More travel and overnight • Will lose customers of long standing — especially close to home

Fig. 2.3   The change from geographical to industrial specialization.

When industry specialization was initially announced for field sales, the salesmen said that they were out of their minds to change. I had some questions and also saw some pluses—I knew many intelligent people had worked on it, and I gave the benefit of the doubt to the program. After the initial reaction, they just started traveling different routes. I thought specialization would hinder me in doing my job because it would spread the men much thinner; they are hard to motivate to make calls, so why put up roadblocks?

Specialization itself has made little impact on us. It raised travel costs from $35 to $40 per man per week. The number of calls settled to the same pattern as before. (It seems to be highly a function of the man.) There is no noticeable change in working with internal marketing and engineering but what is noticeable is deteriorating communications with inside marketing people. I think that this is not because of specialization, but because the job of the internal marketing man is changing—they cannot go into the plant to expedite as much as before.

One reason I was so upset about industry specialization was they were completely switching things around based on reasons in conflict with the real selling situation. They said it would give (1) increased penetration in given industries (but how is that possible when we already have a substantial share of the market in many segments?); and (2) increased emphasis on consumable supply sales. But these don't take aggressive selling. With ink, customers try others, but Printer ink is so good that they come back. With rubber plate, price and delivery is what they want, and we cannot compete with competitors' overnight delivery. When you really analyze the selling situation and why they sell, you find that many reasons are outside the control of the salesman. If that is true, geography should be more important for organization than industry specialization; you can sell more economically.

One thing that is very much on my mind is the increased dependence of the field salesmen on the industry managers that weakens the role of the regional managers. I consciously do things for the fellows to head this off. There is a massive friction between the industry managers and the regional managers. I'm not convinced this is directly a result of industry specialization in the field, but I don't feel the boom in sales is from that either. I'd like to go back to the old geographical organization.

3
THE EVOLUTION OF MATRIX ORGANIZATIONS

INTRODUCTION

Most of today's successful and mature matrix organizations moved to this form through a number of evolutionary phases rather than by a single dramatic change. In fact, one of the basic lessons to be learned by any organization attempting the matrix is that *a successful matrix must be grown* instead of installed. Too many organizations have tried to plug the matrix idea into an existing conventional structure like a new electrical appliance and have blown some fuses in the process.

Which path a particular organization follows depends on its own history and starting place—one path is not therefore superior to another although one path might be more appropriate for a given organization. Furthermore, some organizations move a step or two toward a matrix and then hold with success at some midpoint. Some others skip a phase or two and still succeed. The important point, however, is that the particular path an organization takes, and how, and how quickly it covers that path, must evolve out of the conditions that are specific and appropriate for that particular organization. An all too common mistake is for an organization to look admiringly at a successful competitor that happens to be using the matrix; to see in the other's matrix the answer to its own organizational problems; and then to try and imitate or copy the same in its own. We can almost guarantee failure; not because a matrix would have been inappropriate, nor because it would not yield the desired results once established. Experience suggests that the failure will occur in the inability to evolve along a path in a manner which will be accepted by the relevant members of the organization. The transplanted variety of matrix is rarely accepted, whereas the home-grown variety often thrives.

When imitations or transplants occur and fail, the matrix naturally gets a bad name. It becomes a negative symbol, and to some a "nightmare" and an "abomination." Chapter 5 touches on some of these personal reactions, and Chapter 6 discusses how difficult, if not impossible, it is to resuscitate a collapsed matrix. If the potential of the matrix is to be used and not abused, therefore, we need to understand how successful matrices evolve.

In Chapter 2 we held that there are identifiable conditions which must be present in order for the matrix model to be used appropriately. In this chapter we will hold that there are a number of identifiable phases through which organizations may pass on their evolution into, and sometimes out of, the matrix. We will describe the more frequently used paths that we have observed, together with some of the critical developments that take place in the establishment of a matrix.

As we go through these evolutionary phases, it is difficult to suggest any definitive timetable, although some parameters are likely. Basic and major changes in organization do not often occur more than once in a decade, and the matrix is no exception. Although many organizations have thrived on the matrix for several decades, ten years does seem to represent a minimum life cycle for those organizations that evolve into and then out of the form, and who retrospectively deem the total experience as having been a success. During the matrix life cycle we would point to four somewhat critical times: *decision, installation, institutionalization,* and *exit.*

The first important time is when the concept of the matrix is developed but as yet uninstalled. This is when decisions are made, though not implemented. The appropriate time span for this gestation period should be neither too long nor too short; probably five to ten months. Anything less and an ill-conceived imitation is the likely result. Anything more and the matrix will be stillborn, politicized and talked to death.

The second eventful time is when the matrix is initiated into the existing organization. The key here is to remember that implementation is essentially a process, not a structure. The successful introduction of a well-conceptualized matrix will probably take between one and two years. In more complex organizations, such as in multinational conglomerates, this may take three or four years.

If the matrix is still alive and well at the end of this period, the next critical time is when the understood and accepted matrix is institutionalized—made "permanent." This will probably require adjustments and fine tuning within the matrix framework, but what makes it a critical time is that the matrix must no longer be identified as the child of the chief executive, or other specific top leaders. The idea must hold its weight across managerial generations. If we assume the average tenure of chief executive in United States corporations is between five and ten years, this is a convenient, even though arbitrary, marker.

Assuming the matrix survives all these critical times, there may occur one final passage: knowing when to discard the matrix. Many organizations may never reach this point, and certainly no date can be fixed in its regard. But it is just as important to recognize an idea whose time has passed, as one whose time has just arrived. As we will discuss below, and again in Chapter 9, the matrix is not the ultimate form of organization. Because of the paradox built into dual command, the matrix structure may be the necessary yet never the preferred form. After, say, a decade of working in a matrix organization, therefore, the structure may be abandoned, while the dual systems for processing information and the managerial behavior that works well in interdependent settings may continue on. Matrix, like any other form of established management and organization, may become entrenched and outlive its usefulness.

PHASES AND PATHS OF DEVELOPMENT

There are no fixed number of phases in the evolution of a matrix, but we believe that there are distinct passages and that they do seem to occur in sequence in the larger number of instances. They also seem to parallel, but are not the same as, the critical points in time described above.

Phase I: Beginnings

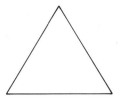

This is the traditional or conventional form of management and organization, in the sense that it is shaped around a unity of command. We begin here because the matrix is shaped from the principles on which this form is based. In this phase, all subordinates have a single boss, though not the same one, and bosses delegate authority and responsibility in equal doses. These latter benefices are distributed throughout the hierarchy according to the scalar principle: the higher you go in the pyramid, the more you have of both. Structurally, management must rank the importance of the alternative organizing dimensions: functions, products/services, geographic areas, client markets. One must be chosen as paramount, and the others are sequenced into one or another of the now divided parts at lower levels in the hierarchy. There is only one basis for differentiating activities, the others are secondary and tertiary.

The classical problem with this classical structure, of course, is that once you choose one basis for your design, you lose the advantages that another choice would have given you. If you differentiate your activities according to functional tasks, you do not have the advantages gained by coordinating around product groupings. If you organize around geographic centers, you duplicate your resources in each location and lose the benefits of scale. And so on. Any design using the unity of command model, whose structural form is the pyramid, results in a trade-off between simultaneously important dimensions.

Phase II: Temporary Overlay

When these organizing dimensions are simultaneously important, but not equally so, a second phase may develop which we call a temporary overlay. The traditional pyramid is kept and there is one major organizing dimension. A secondary way of organizing is created, however, to complement rather than to share or displace the power and purpose of the first. This is the evolutionary route that the aeorspace industry traveled and that the public most frequenty identifies with the matrix idea.

Here, a traditional functional organization is given a temporary overlay in the form of project management. Projects need very close monitoring because incentives and penalties are frequently built into the tightly drawn cost, schedule, and performance requirements. Projects often have rigid delivery requirements which make the normal up-and-down communication channels of functional organization too ponderous and time consuming. And most of all, projects are usually at the frontiers of their particular technology and come burdened with an accompaning high degree of complexity.

Project organization offers a way out of this considerable complexity. Decision making can be focused through the project leader, where the relevant knowledge can be assembled and where activities can be closely monitored. Project managers become "mini-general managers" because they acquire and assemble the relevant resources, because they plan, organize, and control the tasks and activities, and because they take complete responsibility for results.

Most projects, however, are not large enough to contain their own resources. Furthermore, each project is unique and moves through different stages which require different resource capabilities. The need to share resources is evident. The method of doing so is sometimes referred to as decentralized project support. It means that support can come from a variety of sources, including functional units within the organization, outside consultants, subcontractors, suppliers, and customers. At the same time, projects succeed because they are under tight central planning and control, which often use a range of techniques to accomplish control. PERT, CPM, and a variety of network and control methods frequently go hand-in-hand with project management systems and discussion of these techniques tends to dominate the project-management literature.

Conflict is frequently built into the differences between resource groups serving on the same project. Integration sometimes via control systems, sometimes through physical arrangements, and sometimes just through the role of the project manager serves as a way to resolve the conflicting perspectives of the different resource functional groups. These bases of project organization—the project manager, decentralized project support, and centralized planning and control—offer a viable alternative to functional organization, particularly when tasks or projects are complex and of a temporary nature.

Although project management is the most common form of temporary overlay on the basic organization design, it is not the only one. Although not limited to high technology industries, particularly aerospace, it tends to predominate there. In firms and industries that are more marketing-oriented, however, the same principle may be, and often is, applied. A new product or busines development task force, may be formed to work together from the conception of an idea through its development and launching into the market. For the products that succeed in establishing themselves, the taskforce might become a permanent unit (Phases III or IV) or the activity might be handed over to the appropriate product line management; and those that do not succeed are simply disbanded. Temporary overlays may also be formed to coordinate activities around geographic concerns, such as an expropriation threat or a plant relocation.

Phase III: Permanent Overlay

At one time or another almost every organization avails itself of project management techniques to manage temporary, nonroutine tasks. Some organizations make this overlay permanent. The task force may be converted into a permanent team, the project manager may become a product manager. In this arrangement there is still a clear primary organizing dimension, and the overlay is secondary. The brand manager in a consumer-marketing firm is a good example of a permanent overlay.

Because of the temporary nature of program or project management—completing a task within cost, schedule, and performance targets—the objective of the project manager is to go out of business. Product management is oriented in an almost opposite direction. It aims to take an opportunity—an idea, a service, a technology, a product, or a brand—and make it as profitable, extensive and long-lived as possible. The objectives of the permanent overlay manager are to turn potential into an ongoing business, and then to manage that activity without the formal authority that is granted to the formal (usually functional) line managers.

Brand managers, for example, are commonly found as a permanent overlay on the functional organization in soap, food, toiletries, and chemical industries. The brand manager's role is to create and implement product strategies and plans, monitor the results, and take corrective actions. According to Philip Kotler, this responsibility breaks down into six specific tasks:[1]

- Developing a long-range growth and competitive strategy for the product.
- Preparing an annual marketing plan and sales forecast.
- Working with advertising and merchandising agencies to develop copy, programs, and campaigns.
- Stimulating interest in and support of the product among salespeople and distributors.
- Gathering continuous intelligence on the product's performance, customer and dealer attitudes, and new problems and opportunities.
- Initiating product improvements to meet changing market needs.

Its major advantages are that it integrates the various activities involving a product, enables quick response to changing conditions, ensures that minor brands are not neglected, and serves as a visible training ground for promising young executives. At the same time, however, the brand manager is essentially a proposer and persuader, not a decider nor operating manager. Depending on personnel skills, and on the perceptions of others, the manager is either a minipresident or a low-level coordinator.

Permanent overlays also occur in many other forms. An industrial product manager, in contrast to a consumer product manager, has a different focus. Although cutting across the functionally specialized lines, this manager concentrates more on the technical aspects of the product than on marketing considerations. Within the company, the manager spends more time with laboratory and engineering personnel than with advertising and sales promotion, and on the outside spends more time with major customers than does the consumer brand manager. These individuals or groups go by different names in different companies: for examples, Dow Corning has "Business Managers," Texas Instruments has "Product/Customer Centers," and   service firms often have account managers. Other types of permanent overlays may involve area coordinating committees across product divisions (see Chapter 8); and client–market coordinating teams (consumer, industrial, government) across product or territorial groups.

What distinguishes this phase from the previous one is that the overlay is permanent; what distinguishes it from the next phase is that it is a complementary organizing dimension, rather than an equal partner.

Phase IV: Mature Matrix

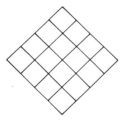

This is the full matrix phase. The essential distinction is, as we defined it earlier, a dual authority relationship. Power is balanced and shared between 2-bosses each representing a different dimension along which the corporation is organized. The essential point is not that power is divided up, but that any unit of power is held simultaneously by 2-bosses. The corollary is that matrix managers, those directly embedded in the core of the form, have 2-bosses. The majority of the book is an examination of this phase.

High technology firms employing the matrix frequently move directly from Phase II to Phase IV in the evolution of their organization. In other words, as the second organizing dimension evolves, it gains equal power and it shares authority simultaneously. Moreover, the structure may be permanent whereas the deployment of people throughout the matrix may be changing constantly. In the aerospace industry, for example, people are changing projects and shifting emphasis in their simultaneous assignment to both their functional bases and project managers.

There are many possible organizing dimensions for the matrix in Phase IV. In domestic organizations, the function–product balance is the most common; and in the global organization (see Chapter 8) the area–product balance is the one most used. As we said of Phase I, other choices may be to differentiate in terms of markets (rural, urban; underdeveloped, developing, developed), or clients (consumer, industrial, government). Service organizations frequently use an area–activity balance, as described in the examples in Chapter 7. Organizations in high technology industries, such as Texas Instruments where there is constant new product development and short product life cycles, have even used time as an organizing dimension.

Phase V: Beyond the Matrix

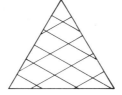

Many matrices continue to evolve beyond this fourth phase. Some begin to see the matrix as a transitional form, and having lived through a full life cycle, they begin to evolve out of their mature matrix. Others find the balance of power continuing to rotate, so that the second, or added, dimension gains primacy. Still other organizations attempt to maintain the benefits of a matrix organization without the difficulties of the matrix structure. We will save our discussion of these variations, such as Phase V, for Chapter 9.

Matrix Is a Verb

Before we move on, however, another word about structure is important. It is a generally accepted notion that as the environment changes, businesses adapt their strategies to meet the new conditions. It is also understood that each business should organize around its particular strategy, and that no one form is adequate for all needs. So we have a progression of change triggering change: environment→business strategy→organization design. While environmental and strategic changes tend to evolve, however, the structure of organizations tends to be static. Structure changes in discrete quantum jumps. This means that while strategy is continually evolving, structure is condemned to a catch up game which it can never quite win. For example, firms implement their strategic plan moving from time 1 to time 2. By the time they reach time 2, the vast majority will still have a structure that was appropriate for time 1. In other words, structures are basically *reactive*.

There are two ways to remedy this. One is to do organization planning together with strategic planning: "If this is what our business will be like when we achieve our strategic plan, what kind of organization will it take to run this new business effectively?" This is a *proactive* approach to organization design, one that is built directly into the strategic process. The other remedy is to view structure itself as a process. Frequent small changes, for example, are likely to be less disruptive than infrequent major shakeups. By analogy, scientists are finding that the best way to prevent major earthquakes is to trigger constant microquakes. The pyramid is the architectural image for the 1-boss model and, in fact, it is one of the most enduring designs built by humans. But its strength is stability, not flexibility. For the latter, the matrix of the 2-boss model is better seen through the imagery of a fabric; able to adapt to stress and change. In the successful matrix, people speak less of the "matrix structure" and more of the process. In other words, *matrix is a verb*.

Because we focused on the structural elements in describing the first four evolutionary phases of the matrix, in the remainder of this chapter we will examine the key roles in the shifting balance of power, the changing mechanisms for data and word processing, and the importance of an institution's intangibles—its culture.

KEY ROLES

There are three key roles in the matrix: (1) the top leadership; (2) matrix managers, who must share subordinates with other matrix managers; and (3) the subordinate manager with 2-bosses. They are rather standard with regard to their form, position, and nomenclature. What makes them different and important is the necessary behavior that is called for if they are to be played appropriately.

Top Leadership

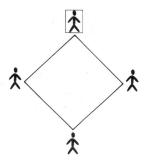

The top leadership is literally atop, or outside of, the matrix. This is not generally appreciated. Even in matrix organizations, the top executives are not *in* the matrix. While they are not in it, however, they are *of* it: it is the top leadership who oversee and guarantee the balance of power. In a corporatewide matrix this is the chief executive and/or the few other key individuals; in a product group or a division matrix it is the senior manager. This individual does not share power with others and there is no unequal separation of authority and responsibility. Formally, the role itself is the same as in any traditional organization. What distinguishes it from the traditional top slot is the leadership process that is applied to the people in the next levels down.

The top leader is the one who must "buy" the matrix. The leader must be convinced of its merits to the point of believing it is the best (not necessarily or likely the ideal) of all alternative designs. The leader must also "sell" it, be very vocal and articulate in developing the concept and enthusiasm among the ranks for it.

Strategy is always set at the top and implemented at lower levels in an organization. In the matrix this means that strategy is fixed by those above the balance of power, i.e., by the one(s) who hold(s) the balance. While different managers below may push for competing strategies, this does not imply that all strategies are of equal power and priority. To make the matrix effective, top leadership has to be at the apex, bringing together the dispersed power. In poorly functioning matrices, the failure to formulate strategy may arise because there is no voice more equal than others.

One of the several paradoxes of the matrix, then, is that it requires a strong unity of command at the top to ensure a balance of power at the next level down. In some senses this is the benevolent dictator: "You will enjoy democracy (shared power) and I will enjoy autocracy (ultimate power)"; or, "I'm OK, you're OK; but I'm still the boss."

Balancing power as a top leader, therefore, calls for a blend of autocratic and participative leadership styles. A clear example of this comes from Mr. Bastien Hello, head of the B-1 Bomber division at Rockwell International. The *New York Times* called his project the most costly and complex plane project in history. In an interview with them, he said,

> Today I have some formidable people working for me. When you have a group like that, you have two choices, running a Captain Bligh operation, or a Mr. Roberts operation. I would call one autocratic, the other group therapy.
>
> If I have to lean in one direction, I would shave a little closer to group therapy. It's not because I, and the fellows who work for me, don't have autocratic tendencies: we do. But if you're going to keep everbody working in the same direction, you've got to have group participation in the decisions.
>
> So I like to get my team of managers together and thrash out problems with them, and I like to hear all sides. It's not that I'm a goodie goodie about it; there *is* malice of forethought to it.
>
> Once they have participated in, and agreed to, the decision, you can hold their noses right to it. It's not that I like group sessions—I don't, they're painful—but they do bring the team along. And once you get them signed up, *then* you can become autocratic about it.

Finally, top leadership plays a specific role in making the matrix succeed because it is the only place where the balance can be guaranteed. Top management will decide whether a single chain of command, an overlay, a balance, or a shift from one line to another will prevail. Again, in classical terms, the executive in the top role will arbitrate disputes that cannot be resolved along the dual lines, and will watch to ensure that no line is so weak that it cannot function effectively vis-à-vis the other.

The Matrix Bosses

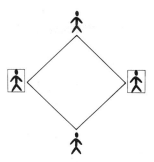

These are the bosses who share subordinate(s) in common with other bosses. These bosses report in a direct line to the top, but do not have a complete line of command below. This is a senior management role. The holder of the role is in charge of an entire function, product, area, business, or service, but is not in full command over the individuals reporting to him or her. These bosses are definitely in the matrix. They share power with equals, often over the same subordinates and usually over information and issues. They are the recipients of the unequal distribution of authority and responsibility. The boss who shares subordinates with other bosses is asked to represent a major portion of an organization's activities, at the same time to take an institutional perspective—the corporate point of view.

Since the heir to the chief executive office is likely to come from this rank, there is generally a great, though diplomatic, battle going on for supremacy among the shared-subordinate bosses. The statesman's posture is an essential ingredient to success. The appearance of being threatened by sharing subordinates is fatal: this is not top-management material nor behavior. Top leadership often uses the matrix to let the candidates for the top spar with each other in a constructive arena. The matrix is a better form than the pyramid is for testing managers abilities to "make happen" by the strength of their personalities, perceptions, and abilities to lead rather than because of their position in the hierarchy.

This boss is aware that subordinates have other voices to attend to, other masters to please. Orders that seem irrational or unfair can more easily be circumvented under the protection of the other boss, than in a single chain of command. More care, therefore, is given to making clear the logic and importance of a directive.

For senior managers who must share their people with other senior managers, the matrix is both a training ground for how to assume the institutional reins and an incentive for no longer having to share those reins quite so much.

The 2-Boss Managers

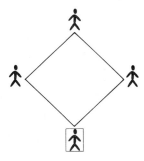

The rule to success in this role is to accept the fact that, while it can place contradictory demands on people, it is the best solution to accommodate simultaneous competing demands. Assume that there is no one best way to organize; each alternative has equally important claims, and the correct choice is both—in varing proportions. Remember that this manager is also at the apex of his or her own pyramid—subordinates to this role need not be shared. It is the multiple demands above and beyond one's immediate command that must be managed. But this manager's formulation is not different from the one for the top role: both must pay heed to competing demands, make trade-offs, and manage conflict that cannot be resolved. Any skillful politician knows that alternative sources of power increase one's flexibility. It is the nonimaginative 2-boss manager who would trade extra degrees of freedom for finite and singular sources of action.

One operating manual for this role, developed after about a year's experience in a matrix, included the following points as practices for managing matrix relationships.

- Lobby actively with relevant 2-boss counterparts and with your matrix bosses to win support before the event.

- Understand the other side's position in order to determine where trade-offs can be negotiated; understand where your objectives overlap.

- Avoid absolutes.

- Negotiate to win support on key issues that are critical to accomplishing your goals; try to yield only on the less critical points.

- Maintain frequent contact with the top leadership, to avoid surprises.

- Assume an active leadership role in all committees and use this to educate other matrix players; share information/help interpret.

- Prepare more thoroughly before entering any key negotiation than you would in nonmatrix situations, and use third-party experts more than normally.

- Strike bilateral agreements prior to meetings to disarm potential opponents

- Emphasize and play on the supportive role that each of your matrix bosses can provide for the other.

- If all else fails:
    you can consider escalation (going up another level to the
        boss in common);
    you can threaten escalation;
    you can escalate.

- Before traveling this road, however, consider your timing. How much testing and negotiating should be done before calling for senior support? Does the top leadership want to be involved? When will they support and encourage your approach? Does escalation represent failure?

This kind of advice relies on managerial behavior, not on organization structure, for success. It sees power derived from personal style and influence as more important than power derived from either position or specialized knowledge. Success flows from facilitating decisions more often than from making them. To remain flexible in this managerial role, it suggests, you must minimize the formal elements: move from fixture to actor, from bureaucracy to process.

All three roles are essential to an effective matrix, where they are present and successful we speak of effective matrix behavior.

SUPPORTING PROCESSES AND PROCEDURES

In addition to a balanced structure and shared roles, a matrix organization should have mechanisms for processing information along overlapping dimensions simultaneously. In a product–area matrix, for example, each product line and each area should prepare separate strategic and operating plans; control systems should account for each dimension separately along and across the overlapping lines of the organizational grid; and all resources should be sought and handled in a similar, dual manner. Requirements for capital, materials, and people are all planned, managed, and controlled according to the needs of each dimension separately. Trade-offs then can be made, if necessary, on the basis of more complete information by the top leadership and its staff outside the matrix. In the human resources function, for example, both dimensions of the matrix should be involved in career coaching and development, in performance appraisal and salary review.

Setting up the formal systems for dual information processing is probably the easiest part of establishing a matrix organization. The difficulties, if they exist, are likely to appear in training individuals to prepare the information and apply the conclusions drawn from the systems. This problem obviously is not unique to matrix organizations. Cultures are also unique to matrix organizations, although matrix organizations often do have unique cultures.

CULTURAL PATTERNS

Cultures are generally spoken of in regard to nations, not organizations, but they are applicable to both. "Top management strategies are not mechanical," writes Peter Drucker[2] "they are, above all, cultural." The term refers to the habits and behavior of a given people in a given period. It describes their values and manners, the characteristic ways and patterns by which they act. By having a shared set of values, individuals develop strong and predictable ways of behaving that identify them with one another. The unique pattern of a culture also differentiates its members from others who do not share a similar orientation or view of the world.

All organizations have to decide how to differentiate their activities and also how to coordinate them around a shared person. Cultural patterns in different societies, however, will lead to different solutions to these common problems. In this regard, it may be said that the matrix is particularly, although not uniquely, reflective of United States cultural patterns. Although matrix is being used more frequently than before in many countries, organizations in other countries have often taken different approaches to similar concerns in ways that reflect their culture.

In German organizations, for example, the division of labor and its coordination are managed in ways that are specifically related to the culture and expected forms of behavior. Top management is comprised of a boardlike group, the *vorstand*. Each member is nominally in charge of a specific functional line, but the group acts as a whole, making all decisions as a collegial body. This is not a matrix as we have described it in terms that reflect the United States culture, but it reflects another culture's adaptation to similar necessities. The *vorstand* diffuses decision making among its members without dispersing it to a larger number of people as in a decentralized divisional mode. Specialization is maintained at the very top in the *vorstand,* which at the same time acts as the most important coordinating body. Where United States culture leads management to a balance of power model to meet its needs for differentiating activities, the German culture creates collegial management to bridge specialized lines.

An element in French culture, and the organizational response in France, also helps us to understand the foundation of the matrix in United States cultural patterns. A cherished tenet of the proper use of power for us has been that authority and responsibility should not be separated, but rather tied together in equal amounts. Our beliefs tell us that this will help guard against abuses of power; it is a guarantor of individual liberty and organizational efficiency at the same time. The French, by contrast, arrive at a different solution for much the same philosophical or cultural reasons. As a guard against absolutist power they have an implicit distrust of any necessary equality and connection between authority and responsibility. As Crozier, the leading French commentator on organizations, puts it: "The French bureaucratic system of organization is the perfect solution to the basic dilemma of Frenchmen about authority. They cannot bear the omnipotent authority which they feel is indispensable if any kind of cooperative activity is to succeed."[3] So, according to Crozier they grant absolutism to the ruler at the top, but then isolate the ruler from imposing an arbitrary will on those below through a body of impersonal rules that eliminate dependency. The matrix separates authority from responsibility at some points in the structure, and United States cultural patterns lead us to worry about how this will affect the independence of individuals and rationality of collective action. The French similarly separate authority (vested in the ruler) from responsibility (vested in the rules), but they do so *in order* to protect individual and collective action simultaneously.

Of all Europeans, the British have the most common culture with ours, and yet the matrix is not very popular there. The reason, we believe, goes to the heart of the two cultures. The British place enormous emphasis on form. Americans eschew form and instead prefer to "let it happen" rather than to do what is "proper." Form for the British, rigid adherence to accepted rules of behavior, is a guarantor of both stable social order and individual liberty. Playing by the rules of the game establishes and protects both privilege and decency, without contradiction. Any organization that is purposefully built around contradictory claims, and therefore provokes conflict that must be managed, does not sit well with the British style. We are reminded of an instance that epitomizes the difference. A British executive was in an American training course. During a session on managing conflict and change, the Englishman remarked, "I do marvel at the way you 'let it all hang out.' That's definitely not our cup of tea!"

Patterns of organization often run in parallel for nations and for corporations. Take, for example, the simultaneous concern for specialized and immediate responses (the push to decentralize) and for coordinated response of disparate efforts (the push to centralize). This description of independent subunits and a struggle to unite them within an encompassing framework guided by a central policy is as equally valid for China before and after the communist takeover as it is for General Motors in the 1920s–1930s. Similarly, the topdown control of subunits, and the need to give them freedom for rapid response based on specialized (technical or local) knowledge, is equally valid for Russia before and after the revolution as it is for the Ford Motor Company in the 1930s–1950s.[4]

The Soviets, in fact, offer an excellent example of how cultural patterns influence the character of organizational solutions. A major change in the governments' economic structure will illustrate the point. Until 1957, a typical enterprise was under the administrative jurisdiction of a ministry, generally in Moscow, and these ministries were organized along industry lines such as steel or chemicals. The industrial ministries were abolished in 1957, and the basis of the reform was to eliminate ministerial "autarchy," that is favoritism, exclusivity, and duplication. They were replaced by regional Councils of the National Economy which grouped all industries within a given region together. New patterns of favoritism quickly developed, and a more costly regional autarchy replaced ministerial autarchy. In 1965 the Soviets reverted to industry lines. The autocratic behavior is deeply embedded in Russian cultural patterns; the structural dilemma, of organizing by activity or by geography, is universal for governments and corporations. The cultural pattern made the organizational choice an either-or one, it ensured that neither trade-off would work well, and it also demonstrated that the effectiveness of a structural response requires a cultural pattern that will support it.

We cannot close this brief look at national culture patterns and their influence on organization design without mentioning the Japanese. Whenever the matrix is discussed with an international group of managers, it is invariably the Japanese managers who feel most at home with the concept and the behavioral style that is called for. Many aspects of Japanese culture are ideally suited to the matrix format: emphasis on the harmony of group relations, commitment to and training for an organization rather than for a specialized role, the subordination of delegated authority to the art of group decision making, and the diffused sense of responsibility of each for the actions of others, to name just a few.

If the Japanese culture is so receptive to the tenets of the matrix, why haven't more of their corporations used it? In part, many have, such as in project teams and task forces. But more to the point, the Japanese don't have to create a formal matrix structure and name for "it" the way we do, because matrix structure and behavior is an intrinsic part of their way of being. Moreover, their cultural patterns underscore the major point of this section: patterns of behavior and a culture that support the matrix are not only essential to make the structure work but they also make the formal structure secondary in importance. The processes *are* the essence of the form in Japan, as they are in successful matrix organization in the United States.

Another parallel between the matrix and the Japanese form of management involves the distinction between making decisions and implementing them. American managers frequently complain that it takes longer to make decisions in a matrix than it does in the pyramid form of organization (see Chapter 6). Americans doing business in Japan often make the same observation concerning how long it takes for the Japanese to make decisions. In the latter instance, however, they also marvel at how fast decisions get implemented once they are made; much faster than in United States organizations, and the total time elapsed may well be a bit shorter. The same point can be made for the matrix: decisions get implemented more effectively, even though they may not be made as rapidly.

The TRW case study at the end of the next chapter is a good illustration of cultural patterns at work in an organization. The description is unmistakably of United States patterns, both cultural and industrial: temporary, fluctuating, complex and interdependent tasks, relationships and structure. It also seems consistent that these characteristics are set in the southern California culture which emphasizes change and a flexible life-style, whereas the TRW headquarters and its divisions in Cleveland have very different needs and patterns. There, the product lines are in more mature businesses, with less complexity, slower change and fewer creative requirements. Following the parallel, the people are more formal and less mobile, the organization structures call for less frequent modification and interdependencies, they are more stable and/or bureaucratic.

Does this mean that there is such a thing as a matrix culture in the organization? No. What we are suggesting, however, is that some organizations have cultures that are most hospitable to a matrix than are others. If there is a strong tradition of rigid bureaucracy, minimal contact even when called for, a belief in the sanctity of reporting lines, etc., then even if the nature of the organization's activity calls for a matrix, it is unlikely that one would be successful unless the culture of the organization was also changed.

What elements in an organization's culture are responsive to the requirements of a matrix? Perhaps the most important is an open and flexible attitude with regards to "the way things are done around here" and to change in general. A tradition of change is a helpful orientation. Open and frequent exchange of ideas and positions on issues is another important feature that seems to be present in functioning matrices. A third characteristic seems to be a shared belief and excitement that those in the matrix are participating in an experiment. They seem to feel they are creating and purposefully evolving their organization, they are not certain where it will lead, but they have some confidence that they're headed in the right direction, and isn't it "really more exciting than fitting ourselves into some fixed arrangement." The sense of challenge and experimentation with new ways to organize and manage also seems to give those involved a sense that they are rather different; not better, but somehow a bit more unique. In a few instances this comes across as pride, but in most cases we have seen it is simply an awareness that "we do things differently" and with it just a touch of satisfaction in that feeling or fact.

We have seen a number of firms do extensive "office landscaping" at the same time that they evolved their matrix. Long corridors, with boxlike offices leading off of them, are an architectural reflection of a bureaucratic pyramid and a traditional organization chart. They channel vision quite literally along narrow lines and hence influence interactions in controlled and restrictive ways. Office landscaping literally tears down walls with the intention of doing the same figuratively, that is, in order to facilitate open and frequent exchanges with a minimum of hierarchy and formality. This is not the open bullpen with row upon row of clerks' desks. Space is rearranged into "family groupings" where those who need to communicate most with one another are in clear view, but where boundaries and privacy are present in the forms of plants, artistic partitions, tables, chairs and cabinet arrangements. The emphasis is on flexibility, informality, and minimization of hierarchy. The physical setting is a reflection of the internal culture.

These examples are not found exclusively in organizations that have a matrix, nor do all organizations with a matrix have these characteristics; but they are indicative of a culture that enhances the necessary managerial behavior.

The case study that follows gives a sense of how these various elements we outlined actually grow inside a particular organization. In the next chapter we will take a closer look at the structural forms, key roles, and processes that characterize a well-developed matrix.

CASE STUDY: CANADIAN MARCONi COMPANY

On the fifth of April 1968, the employees of the Avionics Division of Canadian Marconi Company in Montreal, Canada, were advised that the division would undergo a significant organization change. Mr. K.C.M. Glegg, General Manager of the Division, posted a notice which described some of the changes Avionics anticipated as it reorganized to a matrix:

Managing Change

The general aproach to our organization is intended to reflect the following three features:

1. *Most of our work is made up of large separable tasks having a unique character with regard to customers, equipment, schedules, contractual conditions, test requirements, and so on. This is our program activity and requires program management.* The program manager breaks the main task into subtasks to which other people are assigned, and so on, until a complete program team is evolved.

2. *A relatively small fraction of our work, somewhere between 10 and 20 percent, cannot usefully be viewed as having a unique program character. This is usually initiation and follow-on activity and needs special provision.* Examples are redesign of small areas of a product; updating of a manual; production of spares for an inactive program; and early study and experimental phases of new products.

3. *Most of our work requires similar facilities: selling, design, contracts, manufacturing, purchasing, quality control, scheduling and planning, pricing, and so on. These are functional activities and require functional management.* The total capacity required in any one facility far exceeds that required for any one program, or follow-on activity. This "facility management" or "functional management" is clearly complementary to "program management."

> *Details of Organization*
>
> Using the foregoing discussion and nomenclature, the various groups can be regarded as falling into the following two categories:
>
> 1. *Functional (facility) groups*   2. *Functional and program groups*
>
> | | |
> |---|---|
> | *Mechanical design* | Products and programs group I |
> | *Procurement* | Products and programs group II |
> | *Manufacturing support* | Product support and programs |
> | *Assembly and components* | |
> | *Marketing* | |
> | *Quality control* | |
>
> These groups will contain all program activity and, in addition, a certain amount of functional activity.
>
> The following charts in Figs. 3.1 and 3.2 show the organization of the division and of a typical program.

Predeterminates of Change

Keith Glegg had worked his way up through the technical arm of Canadian Marconi to become chief engineer and then General Manager of the Avionics Division (see Fig. 3.1). He was an important inventor and innovator in basic frequency-modulated continuous wave (FM-CW) Doppler radar technology. This FM–CW technology gave Avionics a world leadership position in Doppler radar equipment design and production. All Avionics' equipment designs were at the state of the art of the technology in their field at the time of their design, a result of the importance Glegg attached to the role of innovative research and development engineering for the department's future.

Doppler radars were used in airborne navigation systems to determine aircraft velocity and distance by bouncing microwave signals off the terrain and calculating the desired parameters from the "Doppler effect" of returning signals. A typical Doppler system was composed of several "black boxes" of electronic and electromechanical equipment. The Doppler systems in the Avionics product line sold for prices that ranged between $20,000 and $80,000 per system.

As the division grew and Avionics' success with Doppler systems brought large increases in sales, Glegg's preoccupations became considerably more managerial than technical. He began to reassess some of his own thinking about organizations. The organization appeared too weak, both structurally and managerially, to cope with the increasing complexity of his division's activities.

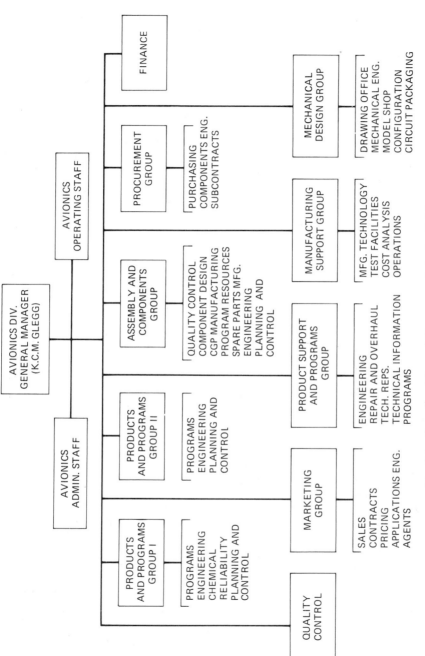

Fig. 3.1 Canadian Marconi Company—Avionics division.

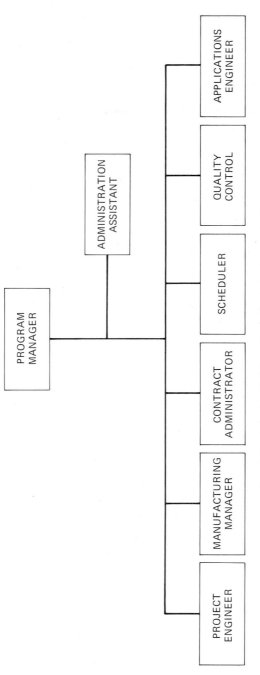

Fig. 3.2 Avionics division—typical program organization.

Glegg was finding it impossible to cope with the number of major decisions that had to be made. Six major programs and several minor ones were in different stages of design and/or production (see Fig. 3.2). All had different customers, sometimes in different countries. Every program's product, although they were all Doppler radar systems, was significantly different from every other one, particularly in its technology. Nevertheless the programs had to share manufacturing facilities, major items of capital equipment, and specialized functions. Glegg felt he had to find some way to force the whole decision process down to some level below his own. He described some of his thinking.

> Even in all the difficulties that we found ourselves in at the time, many of us saw that if we could succeed in managing the place effectively, we could get out of difficulty and into more products and the necessity for still more decisions. I started looking for a way of making the system nonlinear; a way of decoupling the number of decisions that I would have to make from the number of involvements of the division. Eventually one has to decouple the one nervous system from this linear attachment to all the involvements. And the way you do it, is to put another nervous system between you and it.

Glegg identified the uniqueness of the task as one of the key factors determining the particular approach taken to matrix design (see pp. 58–59 above). He felt that in Avionics' business the task could be isolated cleanly and simply. More important, however, was his feeling that the isolation of task was something that could be understood by all the people in the division. After extensive discussion with his subordinates, it was decided that a matrix should be the general form towards which Avionics must move.

Corporate Management Unfreezing . . .

Convincing himself and his managers was one task for Glegg. Convincing corporate management was quite another.

> First of all, I had to get agreement from the corporate management that they would allow me to implement such a scheme. There was a tremendous amount of opposition. There were people opposing it for the usual reasons, such as one man reporting to two people. That's not just against organization–it's against God. It's agin' the Bible.[5] No man should serve two masters. You shouldn't do that. There were people on the accounting side of the house saying, "It's going to cost too much." Obviously, it's a much more expensive way of managing than a straight functional way of managing.

You could break down the objections they were presenting into two classes. One of them was semiphilosophical notions of management and people, that you could overcome simply by talking enough. The practical objections such as, "But it's going to cost more," were more difficult. I guess the way I ended up putting that one out of the way was to say, "Look, what you're telling me is that I'm going to end up with a system that is overmanaged. Now if you ask me, I can tell you what the cost is of this overmanagement. But now I have to ask you a question. What is the cost of undermanaging? I can give you a bounded cost for the over-management. That is, I can count up the pieces and multiply them by their salaries, and I can tell you what the cost of overmanagement is. I can give you absolute upper bound. Now you tell me, what are the costs of undermanagement? Are they $100,000 a year? Are they $1 million a year? What are they?"

Well, at the time we had so many problems of undermanagement on our hands, that it was easy for me to illustrate that the number was almost catastrophic. So I got a kind of grudging understanding that I could go ahead, but it had better work or both it and I would go.

. . . And Unfreezing Everyone Else

Glegg then went about selling matrix to the rest of the organization. A series of meetings for middle managers answered some of the immediate questions, and an extensive notice posted throughout the department was designed to inform the rest of the department members of the intended changes. However, Glegg was aware that neither notices nor speeches could supply the real answer. The people would have to see the system function to have faith in it. He described how he believed the process of acceptance took place for many.

It took the better part of two years to work through enough of the cases to remove the question whether it could work. People are still left with the question as to whether it works for them. If a change has caused them to lose something or other, then it still isn't working for them. But at least over two years, if the thing works in some objective way for most of the people, then it does work.

Establishing the Program Manager's Boundaries

Glegg and his immediate subordinates then faced detailed decisions about the structure. In many matrix organizations the program manager reports directly to the organization's head: the department general manager or divisional vice-president. Should the program managers report to Glegg or should there be a level between him and the program managers? While it was evident that Glegg could have coped with the number of program managers initially envisioned, it was also evident that as their number grew he would be unable to manage both them and the functions. Glegg commented:

> The program manager is the first level at which integration occurs. It is the first level at which the company's business is conducted whole. When he is not available, either because he's sick or traveling or something, it is not practical to go down in the organization to replace him. You have to go up.

Although it was the functional managers that one had to go to when program managers were unavailable, Glegg went to considerable effort to make it clear to the functional managers that the program managers were nevertheless their seniors. They were senior in the sense that they were responsible for "the business whole," and no functional manager enjoyed that privilege. This is also reflected in their salaries.

> We had to decide on the extent to which a program manager would control his facilities and his resources. That is, what would a program manager be? We decided on an arrangement which took us far over on the side where the program manager controls everything. There are some problems with this kind of an arrangement. For example, how do people who are attached to this guy, on an essentially temporary basis, get their progress review done? That's a hard, practical question. The way it was resolved was that the employees progress review would be done jointly by the program manager and by his functional shop. It seemed important that when a person was reporting to two people, that the two people were indeed contributing to the assessment of performance, and that the person with 2-bosses felt it.

Selecting Program Managers

Engineers dominated the Avionics Division. Glegg and six of the eleven group managers held formal engineering degrees. In addition to the very large numbers of engineers within the formal engineering groups, many of the marketing and applications personnel were engineers and a substantial number of the senior personnel in the production and manufacturing areas held engineering degrees. Any consideration of the characteristics needed to be a program manager under the proposed matrix had to face this issue of professional background squarely.

> We wanted to start with an engineer and turn him into a program manager, because the root of the most serious problems is technical. But the object was not to get an engineer and turn him into a program manager so he could solve the problem. The real object of starting with

an engineer was to make certain that there was someone who could assess how to commit resources to the solution; someone who could give order to the problems a program manager was bound to encounter.

There are difficulties inherent in the notion that you're going to retread an engineer. One of the things you have to do is help him suppress a tendency to want to become his own project engineer. You have to help him resign himself to the fact that he's no longer an engineer; but that he's going to have to use what he knows about engineering to run the risk of the generalist.

From Program Management to Product Management

In the late 1960s the division expanded rapidly with the economy, and the aerospace industry enjoyed the tail end of a long economic boom. Mid-1970 sales reached a peak of $36 million and employees numbered in excess of 2000.

However, the year 1970 also ushered in an end to North America's economic boom and the aerospace industry was about the hardest hit of all. Avionics' dependence on programs with fixed termination dates left the division in a very vulnerable position. In some cases, follow-up contracts didn't come. In others, cancellations left design and production areas with little or no work. Avionics needed a different type of business to ensure its survival. It needed products that would endure and applications that went beyond only one narrowly defined program. It also needed an organization that would support such a strategy. Avionics' response was to change the program management form of matrix to a product management form; from an overlay to a more permanent balance. Glegg explained the difference.

The product manager's object in life is almost exactly the opposite of the program manager's. The object of the product manager is to take an opportunity that he has and extend it as far as into the future as he possible can. To start with an idea, a product, someone else's product, and make of it whatever he can; as big, as long-lived, and, of course, as profitable and productive as he can make it.

The object of the program manager is to take a task which is well defined with respect to its schedule, cost and function, and to execute it. A program is really a special case of a product. A program represents a kind of singularity in the product where the intensity of management it needs simply requires that you attach one nervous system to it exclusively.

So far as I can tell from the literature, there's a lot of experience with the program management idea, with the fixed time, fixed money, fixed outcome situation. There is much less experience with the product manager, of the type we have tried putting into existence. What we're finding, however, is that it's an extremely powerful way of getting motivation into the system, because what the product manager has become is a kind of mini-general manager.

Modifying the Control System

To have product managers make the most of their own decisions, Glegg insisted that the accounting system reflect the organization design. With appropriate cost information and profit and loss data available to them, program and product managers could contain decisions at their level. They only had need to approach the group manager level on financial matters when major capital demands were being placed on the division.

Program or product reporting was broken out by total sales and further subdivided to provide as much managerial visibility as possible. Within each category, program and product managers could also determine individual charges by the cost centers which represented the different functional activities. A parallel system recorded functional costs. Every direct charging employee had to be accounted for by a "legitimate" charge number so that only overhead people in the functions could remain unassigned to a particular program. As a consequence, slack personnel in the functions were quickly identified by their charges to program or product numbers.

Communication and Openness

In the early phases of implementation, few people fully understood the matrix concepts. Others understood them hardly at all. Believing that effective and open communication was central to successful implementation, Glegg went about creating an atmosphere to support his beliefs.

> One way of ensuring that the matrix would work was to ensure that functional shops always had the best possible grasp of what the business side of the house was likely to want. It seemed vital that all the resource pieces have the longest possible view of what the demands on them would be if they were to offer the most help.
>
> To do this, I decided that we would have meetings every Friday and talk about all the problems in the division. The people who attend these meetings are group managers in charge of resource functions and group managers in charge of task groups.
>
> At this point, the meetings are as open, frank, relaxed, and productive as one could hope to find in any industrial organization. We trust one another and, maybe what's even more important, they trust one another.

Glegg felt that these meetings lay at the core of successful matrix management. It was a mechanism whereby every part of the organization could question every other part and come to understand better each other's contribution. After approximately two years, one meeting a month was expanded to include program and product managers in addition to the group managers. With the variety of experiences that was building up among the product and program managers, they could learn extensively from each other's problems, successes, and failures. Two years later, this meeting was expanded to the second-level functional managers and a function-by-function critique of each of the areas had commenced. Glegg described the contributions gained from adding the product and program managers to this communication process.

> This allows people to hear from one another and understand that they all have problems; to build a kind of confidence in themselves because they hear about other people's experiences and solutions in similar circumstances.
>
> I'm getting the impression that the communication propagation down through the product manager is more effective than through the functional shop because they are more closely knit to each other.
>
> Holding this clearly delicate "net" together is the most challenging part of my work. The most powerful tool one has, and it is really simple after you get living this way, is openness. I don't have any secrets at all. Anything they want to know, I'll tell them. Furthermore, it's easier for me to know things now, with a lot more activity going on, than it was even two years ago. I don't have to go after information. People come in and tell me things they think I should know. In fact, they'll come in and tell me things in the presence of whoever else is in the room, since we've no more secrets in the division.

Maintaining the Product/Program vs. Function Balance

As a final comment, Glegg talked about the problems of achieving the "right" balance between the two sides of the organization.

> That was a very hard part of living with group managers early in the program. What I had to say to these people who report directly to me was that the people at the next level were, in a sense, organizationally in more control than they were. It was hard to do initially. For a functional manager to derive satisfaction from such a system is very complicated. It's difficult for him to find a way to achieve success because the organization is structured so that its main success indicator is attached to the task and not to the functions.
>
> So he has to derive his pleasure in vicarious ways, and get his satisfaction sideways. They have learned to do that largely by identifying with me.

At first, it was very, very difficult and frustrating for them. This would often lead to impasses since the functional manager would see his so-called prerogatives challenged. The program manager will come into the procurement shop, for instance, and say, "I don't like the way you're buying my stuff," and be answered by, "Look, I'm buying it better than you could." Then the program manager will say, "Well, I'll show you. Just give me two people out of your shop and I'll attach then to the program and buy it myself."

Well, we have gone from that unhappy state of affairs to the situation today where we foresee two months ahead that a program is going to develop a heavy procurement requirement. We'll agree that the functional manager will go the product manager or program manager and say, "Look, you're going to develop a need for a very heavy procurement operation. We can do it in a number of different ways. We can get one or two people, on your job or I can keep them and try to do it for you here—over in the shop. How do you want it done?" We're that extreme. The functional managers long ago ceased worrying about their prerogatives. They see that the real satisfaction will come from allowing people to perform in the most effective way.

Something reciprocal has also happened. The necessity for product managers to go bashing functional managers has largely disappeared. The system has become supportive. I know that we'll all live longer for it.

It's now a real beauty to observe. You can see people anticipating, talking to others about a need that they can see others are going to have, and asking how they can help. The functional shops are now receiving the most unbelievable bouquets from product managers. It's been a long time coming. But the average functional manager today knows now that he performs an important function. And he now knows how to perform it in order to be most useful.

NOTES

1.   Philip Kotler, 1972. *Marketing Management.* Englewood Cliffs, N.J.: Prentice-Hall, pp. 287–288.

2.   Pete Drucker, 1973. *Management.* New York: Harper & Row, p. 746.

3.   Michel Crozier, 1964. *The Bureaucratic Phenomenon,* Chicago: University of Chicago Press, p. 222.

4.   Franz Schurmann, 1969. *Ideology and Organization in Communist China,* Stanford: University of California Press, pp. 298–303.

5.   "No man can serve two masters: for either he will hate the one, and love the other, or else he will hold to the one, and despise the other. Matthew 6:24.

4
THE MATURE MATRIX

INTRODUCTION

In Chapter 3 we examined the several ways organizations can evolve toward the matrix. But what does a matrix organization look like that has arrived at Phase IV? What are the characteristics of a mature healthy matrix? To answer this question with an organizational chart is totally insufficient. We will need to get behind the bare structure. To do so in this chapter we will be addressing three principal questions. What various *forms* does the mature matrix take? What kinds of *processes* are more prevalent in matrix organizations? What kinds of *managerial behavior* characterize the matrix kinds of behavior that differ from those in conventional line organizations? We will address this last question by looking in detail at the behavior of managers in the three key roles that we set forth: the general executive who heads up the matrix, the 2-boss manager at the foot of the matrix, and the matrix bosses in the middle. Each of these managers needs to develop some special skills and attend to some special issues if the mature matrix is to live up to its potential. And if we were pressed to pick one word that characterizes the potential of the mature matrix, it would have to be *flexibility*. The matrix idea offers the potential of achieving the flexibility that is so often missing in conventional single-command organizations and reconciling this flexibility with the coordination and economies of scale that are the historic strengths of large organizations. This chapter will provide both some specifics around this assertion and some tentative explanations.

VARIETIES OF MATURE MATRIX ORGANIZATION FORMS

The matrix idea in its developed phases lends itself to a great many forms. The authors have seen working examples of full and partial matrices, layered matrices, headquarters as well as field matrices, and matrices that move beyond dual command structures to three and even four simultaneous lines of command. We will comment in this section on this variety and some of the reasons for it.

To start with a single example, some organizations have established a matrix within a single functional department while the remainder of the company proceeds to use a conventional chain of command structure. We have seen this most frequently in marketing or engineering departments. In marketing it arises around the role of brand manager whose work cuts across the specialized marketing groups such as promotion, advertising, field sales, and market research. A similar pattern can grow up in engineering areas where the various engineering specialties (mechanical, electrical, computer, etc.) would be regrouped for working assignments under product engineering coordinators. This kind of partial matrix has also been used to organize maintenance activities such as between supervision by trade and by factory floor areas. It has been our experience that as the matrix approach is used with groups performing more routine operations increased care needs to be taken in clarifying role relationships with all parties.

Some organizations have evolved layered or double-decked matrix structures. An example drawn from a medical school can demonstrate this possibility. This medical school had complemented its traditional chain of command built around medical specialty departments by creating a cross-cutting set of program departments: undergraduate, education, graduate education, continuing education, research, etc. The department of medicine, as a large department, found it useful to go one step further and set up a matrix within the department. This broke the departmental faculty into subspecialty groupings (cardiac, renal, gastrointestinal, etc.) and also appointed departmental leaders for each major school program.

Some mature matrix organizations directly involve only a small percentage of the people despite their large size while others involve a high percentage. In the latter category a national CPA firm has a single large matrix that starts at the top with the senior operating partner and moves down through several echelons of a dual functional/geographic structure to the 2-boss professional at the level of the individual CPA contributor in a local office. Only clerical employees are left out of direct participation in the matrix. The medical and CPA examples are discussed in greater detail in Chapter 7.

A very limited number of organizations are operating with three and even four simultaneous chains of command. Texas Instruments, one of the pioneers in organizational innovations, has developed a complex structure which we would describe as a three-dimensional matrix although they do not use matrix terminology. Texas Instruments has matrixed a product structure against a functional structure and, more recently, added a time-oriented structure known as "Objectives, Strategy and Tactics." Their OST program has its own chain of command and its own budget. It is charged with innovation, developing the future business of the firm. Individuals are appointed within this structure as objective managers, strategy managers, and tactics managers in a hierarchy of goals and people. Some of these managers are full time on this assignment while others carry concurrent responsibilities in the functional or product hierarchies. All OST managers are provided with a budget and their progress toward agreed-upon innovation targets are regularly reviewed. Texas Instruments executives are convinced that their impressive corporate growth record would have been unattainable without this system.

Finally, we would call attention to the four-dimensional matrix that has evolved one step at a time at Dow Corning.[1] In addition to the functional, product, and temporal structures used by Texas Instruments, Dow Corning has added geography as a fourth dimension. This, of course, must represent something of a limit. Such a supercomplex structure is mind blowing to most managers and designers of organization charts, but it certainly seems to exist in thriving condition and by all reports meets the test of a structure that serves to draw people into their work rather than a structure that gets between people and their work.

We have cited some of the various forms mature matrices take with no intention of presenting a complete catalogue of such forms. The point of these examples is simply to show that the matrix idea lends itself to many variations. It is a very flexible form of organization, adaptable to varied circumstances. There can be various explanations but one simple and powerful one is that once we leave behind the very restrictive concept of "1 person—1 boss" the possible combinations of organizational forms explode. The abandonment of this restrictive concept by itself seems to excite the imagination of managers to design forms that suit each firm's unique set of circumstances. It forces our minds to explore creative new approaches to organizational design. And once the process is started, it tends to continue. Mature matrix organizations are anything but static. They have a strong tendency to keep changing form in a unending series of small structural shifts. Many of these shifts are carried out at middle management levels or lower, without any significant inputs from top management. Managers in mature matrix structures tend to give up trying to keep any up-to-date precision in their organization charts for this very reason and use only some general schematic charts for educational purposes. In its maturity, matrix provides a very flexible form of organization.

MATURE MATRIX PROCESSES: POWER SHIFTS

The structure in most traditional forms of organization rarely changes. The patterns of political power, authority, and status and the investments in seniority and security are all bound up with a hierarchy that carefully guards career paths and maintains entrenched positions. Resistance to change is everywhere in vertically structured organizations. It is a direct consequence of the repetitive patterns of activities that have been learned and standardized in order to realize the scale economies of bureaucracy. These patterns do not change easily. Efforts to shift power systems trigger a power struggle whose turmoil is costly in human and economic terms. For example, a firm may have been dominated historically by the marketing function. This may have been a very useful arrangement but if industry changes require that more clout be given to research and development, the resistance to power shifts can be costly.

Matrix organizations are similar to traditional organizations in that each of the dual orientations has its own hierarchical structure and within those hierarchies are individuals who also guard their territories. But matrix organizations are also very different. New functional activities, new projects or new products or businesses can be phased in and old ones phased out without resorting to major changes in the basic layout of the organization. And since such changes can take place in a matrix without significant structural revision, at least the trappings and status of authority can be retained even as important shifts in power are resulting from a new or different mix of businesses or resources. Power shifts are never easy but the mature matrix improves the chances that they can be managed without a disruptive power struggle.

MATURE MATRIX PROCESSES: LOCATION SHIFTS

Mature matrix organizations display a degree of flexibility in handling the physical work location of employees that is seldom seen in more conventional organizations. People in matrix organizations become accustomed to moving their work locations on a fairly frequent basis. In matrix organizations the question always arises as to whether functional specialists should move to the site of their assigned business group or should continue to stay grouped with fellow specialists. The answer to this question in the matrix covers the full spectrum from a single business area manager, with nothing more than a small clerical staff sitting adjacent to him or her, to business area groups that contain many representatives of every functional area who are physically located in a building or plant site completely remote from the primary functional groups.

The decision rests once again with what's important in the environment and what is needed to compete in the particular business at issue. If the industry is electronics and the business is the design of equipment for rockets and satellites, then high quality technical solutions are the primary criteria for survival and for success. The physical arrangements here will always honor the integrity of the functional group first. If the industry is still electronics but the business is commercial products (e.g., television sets) where market considerations are key competitive factors, then more resources will be physically located with the product or business group because product–market concerns are the dominant ones for success in that business.

TRW Systems, whose mature matrix is described in the case study at the end of this chapter, has made frequent use of physical proximity for integrative purposes and has referred to it as "co-location." They found co-location to be a necessary device but used it cautiously because they quickly learned that people separated from functional groups for long periods of time "go native" and forget their functional orientation. TRW Systems developed a rule of thumb that after six months to a year a functional person who had been co-located with a project had to return, physically, to the functional group.

MATURE MATRIX PROCESSES: PRODUCT INNOVATION

A significant and not always intended outcome of matrix organization is its apparent capacity to foster product innovation. In those industries where continuous product innovation is a criterion for survival and success, organizations that have adopted the matrix seem to fare well. The following explanations, all of which have some foundation in the literature on the management of research and technology, begin to account for the phenomenon:

- In product or business managers, the structure singles out a "product champion" to take responsibility for an idea, a product or a technological breakthrough. Product managers have a generalist orientation that directs their concern to all aspects of the innovative activity and they frequently have a strong entrepreneurial bent.

- Product or business area teams contain a diverse set of members who bring different contributions to the activity. This diversity is often considered conducive to creativity.

- Product managers are close to both the market and the relevant technology and can develop a good "fit" between the two.

Decision making is centered at a level where the appropriate knowledge and information exists.

- Product teams comprise some of the more elite and talented members of the organization. They frequently hold multiteam membership and can carry over the stimulus and the ideas from one team to another. Many have a high tolerance for ambiguity and good interpersonal skills, characteristics which the matrix tends to self-select, and this enhances a climate of creativity and problem-solving that is conducive to good working relationships and a willingness to go beyond "conventional wisdom."

While these conditions are frequently found in matrix organizations, they also exist in the organization of innovation in many nonmatrix situations. As such, they do not adequately discriminate according to the unique characteristics of matrix organization. A more interesting interpretation is that some or all of these apply, but there is more: the product teams have access to all the benefits and large-scale economies associated with functional organization but are small enough, organic enough, and autonomous enough to still maintain the "innovation characteristics" of small teams. The "critical mass" knowledge for effective innovation exists, but not with its accompanying burden of rigid bureaucratic structure.

This might be better explained with an illustration. Texas Instruments (TI) has been extremely successful with a matrix organization that has over 200 product–customer center (PCC) managers, each of whom manages a small business of his or her own, with long- and short-term responsibility, and with access to some extremely efficient functional groups.[2] The PCC managers are selected from the ranks of engineers who have demonstrated high technical innovation competence. This undoubtedly helps. But more significantly, the former Chairman of TI, Patrick Haggerty, has stated that he can tell when a PCC manager is going to make it as a successful innovative manager by the way the manager learns to utilize the enormous resources that the functional organization can offer.[3] This is another way of defining a "large organization entrepreneur." The comments of a TI vice-president elaborate this philosophy.

As TI has continued to grow, we have sought to preserve the environment in which innovation can continue to flourish at all levels. The difficulty with this approach by itself is that the innovative efforts, even when they exist, tend to come out fitting the size and resources of the decentralized units. . .what we're trying to do at TI is preserve the environment of the decentralized product–customer centers, but at the same time, to knit them tightly together within an overall goal structure.[4]

We see matrix, then, as being able to take advantage of the two orientations that dominate its structural arrangements. On the one hand, the scale economies of large organizations can be achieved, and on the other, the sensitivity, flexibility, and adaptability of small innovative teams is realized. The benefits of standardization are coupled with the capacity to respond to change—a way to have your cake and eat it too. Another way to put this point is that learning becomes a constant condition for some people in the matrix organization. They "learn to learn." Learning is the fusion of differences or variety to develop a new conceptualization or position. This is the phenomenon we see in matrix organizations. When new products or functions are added to the organization, this learning to learn is manifested in the ability to accept these changes without resistance even though the complexity of the organization frequently increases significantly. Where new product innovation is a prerequisite to doing business successfully, learning to learn describes the patterns whereby organizational entrepreneurs optimize the benefits of scale economies with the flexibility and creativity of small size—patterns made possible by matrix.

One theme so far in this chapter has been flexibility, whether in form, in power, in location or in products. The rewards of a mature matrix are certain to be attractive to many managers but they are likely also to seem unreal. How are these general results achieved at the level of individual managers and employees? What managerial role behavior characterizes the matrix that distinguishes it from more conventional organizations? Only if these questions are directly faced can we expect to achieve some sense of reality in regard to the mature matrix. Therefore, let's take a closer look at the three key roles we outlined in Chapter 3 and see how they operate in a mature matrix.

KEY ROLES: THE TOP LEADERSHIP

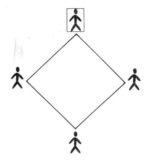

The general executive of the matrix has the unique role of heading up both of its dual command structures. It is our analysis that this role involves three unique aspects; *power balancing, managing the decision context,* and *standard setting.* These three processes, while of concern to any top executive, take on a very special importance in a mature matrix. The reason for this importance is not hard to find. It stems directly from the three basic reasons we have discussed as to why a matrix can be a desirable organizational form. We have seen that the existence of dual pressures called for simultaneous and balanced decision making. The general executive's critical role in achieving such decision making is to establish and sustain a reasonable balance of power between the two arms of the matrix. The second necessary condition calling for the matrix was the very high volume of information that needed to be processed and focused into key decisions. If the organization is to cope with such a load, the top leader must be only one among several key decision makers. The top leader must delegate, but cannot delegate the job of setting the stage for other decision makers. Top leaders must manage the decision context. Finally, they clearly must set the standards of expected performance. Others can and must contribute to this process, but unless the top individual has high expectations for the organization, it is unlikely that the matrix will respond adequately to the environmental pressure for resource redeployment that we have identified as a third necessary condition for a matrix. Each of these three special aspects of the top leader's role in the matrix will be examined in more detail.

Power Balancing

This element of the general executive's role is, in our experience, vital to mature matrix performance. Any general manager has to pay attention to this process but it is uniquely critical in matrix organization. If we contrast the pyramid diagram of a conventional hierarchy to the matrix diamond diagram, we have a clue as to why this is true. The diamond diagram, unlike the pyramid, is inherently unstable. For the structure to stand in the face of environmental pushing and pulling it must be constantly rebalanced with "hands-on" top leadership. The analogy is crude but relevant. Managers in this role are usually quite explicit about this requirement of their job. It needs very regular attention.

The basic methods general executives use to establish a power balance are both obvious and important. The two arms of the matrix are, first of all, described in the formal documents as being of equal power and importance. The top executive tries to use all possible methods to reinforce this message and one method is to establish dual budgeting systems and dual evaluation systems.

Most mature matrix organizations adopt dual budgeting systems: systems in which a complete budget is generated within each arm of the matrix. As with a double-entry accounting system, the dual budgets double count everything, but each in a different way and for a different purpose. Functional budgets are primarily cost budgets (unless the function sells its services outside). They begin with product and business estimates of work required from each functional area, usually in work hours and material requirements. Functional groups then add indirect and overhead costs to these direct hours and return an hourly rate for services to the product or business managers. Products or business units accept these rates or they challenge them, sometimes by threatening to buy outside. When the rates are approved and accepted for all the different functions, the business groups are able to develop profit and loss budgets for each of their product lines.

Budgeting time is the time when the differences in business and function outlook are most evident. Business units, for example, have little sympathy with functional desires to hold people in an overhead category for contingencies or for the development of long-term competence. A business unit is hard pressed to see why it should pay, even indirectly, for the development of competence that may not be needed for several years or where the benefits will accrue primarily to another business when its own business concern is with short-term profit and loss.

The parallel accounting systems provide independent controls that are consistent with the characteristics of the work in each type of unit and which recognize the partial autonomy of each organizational subunit. Each unit has the means to evaluate its own performance and to be evaluated independent of others. One CEO described the dual control systems in his organization as follows:

> The accounting system matches the organization precisely, so that's an aspect the product manager and I don't have to talk about. He can see how he's doing himself. When resources seem to be a problem, then I must get involved.
>
> Both product managers and functional managers get accounting evaluation. The functional shops have budgets but little spending money. They have a cost budget, but in theory it's all released into the projects. From the functional side the accounting system locates and isolates unused capacity. As soon as the task requirement disappears, the excess capacity turns up. The functional shop then has a "social" problem. The key thing is that the excess turns up immediately. There is no place to hide. Matrix is a free organization, but it's a tough organization.

With dual budgets some interesting possibilities arise to achieve a flexibility of organizational response. In the organization mentioned above, the CEO resolved an internal dispute between a product group that was lobbying for control of repair and overhaul contracts on products in the field that it had developed and sold, against the protests of a functional group which had always managed the organization's field repair and overhaul activity. In the resolution, the function remained in charge of the activity but, via the control system, the product group was credited with the lucrative profits from all repair and overhaul contracts on its products. Both sides were satisfied.

Dual personnel evaluation systems go hand in hand with dual budgeting to help sustain a power balance. If a man or woman's work is to be directed by two superiors, both, in all logic, should contribute to that person's evaluation. Occasionally, the duality is nothing more than a product or business group sign-off of an evaluation form prepared by the functional boss. At other times, the initiation comes from the other side, primarily because the individual involved may have been physically located with the product or business unit and had very limited contact with the functional unit over the evaluation period. Irrespective of the particular system's design, however, the person with 2-bosses must know that both have been a part of his or her evaluation if that person is to feel committed to consider both orientations in his or her activities. For this reason many matrix organizations insist that both superiors sit in on the evaluation feedback with the employee and advise the employee of salary changes together so that rewards will not be construed as having been secured from only one side of the matrix.

These basic formal arrangements for setting up a reasonable balance of power are essential in a mature matrix but they are seldom sufficient. Too many events can throw off the balance and this disequilibrium needs to be caught by the general manager or it can evolve into a major power struggle and even an ill-advised move away from the matrix. The matrix can be thrown off balance in many ways but disequilibrium is frequently caused by a temporary crisis on one side of the matrix that is used as an excuse for mobilizing resources in that direction. Up to a point such a reaction to a true crisis is certainly appropriate but it can lead to a lasting imbalance unless corrected by the general manager. A more lasting source of instability arises from the fact that product and business area managers manage a "whole business" and thereby have that special mystique associated with "bottom-line responsibility." This is a source of power. They are seen as the sources of revenue—the ones who make the cash register ring. The general manager needs to be alert to this one-sided source of power to avoid its unbalancing potential. The profit center manager is often tempted to argue that he or she needs to have complete control over all needed resources, but this kind of argument has no place in a matrix.

Given the inherent power instability of the matrix, the general managers of mature matrix organizations use a wide variety of supplemental ways to trim the balance of the matrix on the margin. These methods are not new but are worth remembering as especially relevant for use in a matrix. Five such means follow.

- Pay levels, as an important symbol of power, can be marginally higher on one side of the matrix to be a countervailing force.

- Job titles can be adjusted between the two sides as a balancing item.
- Access to the general manager at meetings and informal occasions is a source of power that can be controlled as a balancing factor.
- Location of offices is a related factor that carries a status or power message.
- Reporting level is a frequently used power-balancing method. For instance, product managers can report up through a second-in-command while functional managers report directly to the general manager.

Managing the Decision Context

There is no substitute in a matrix for the sensitive management of the decision context by the top leadership. The existence of a matrix is an acknowledgment that the executive leadership cannot make all the key decisions in a timely way. There is too much relevant information to be digested and too many different points of view to be taken into account. But the top manager must set the stage for this decision making by others. The top manager must see that it happens.

We have seen already that dual environmental pressures and complexity make conflict inevitable. To cope with this situation, the key manager must sponsor and act as a model of a three-stage decision process. (1) The conflicts must be brought into the open. This is fostered in the matrix structure with its dual arms but beyond this the given manager must reward those who surface the tough topics for open discussion. (2) The conflicting positions must be debated in a spirited and reasoned manner. Relevant lines of argument and appropriate evidence must be presented. The executive manager's personal behavior needs to encourage this in others. (3) The issue must be resolved and a commitment made in a timely fashion. The leader cannot be tolerant of stalling by others or of passing the buck up the line.

All of these decision processes call for a high order of interpersonal skills and a willingness to take risks. They call for a minimum of status differentials from the top to bottom ranks. The top leadership can favorably influence these factors by their own openness to dissent and willingness to listen and debate. One of the noticeable features of most leaders of matrix organizations is the simplicity of their office and the relative informality of their manner and dress. Some of these aspects of matrix management will be dealt with in more detail in Chapter 5, but it needs to be emphasized here that this needed behavior must start at the top. It is all part of setting the decision context. This is well illustrated in the comments of a TRW department manager who was musing on the possible impact of continued growth on the climate of the organization.

> I think the climate here will remain. It's a function of top management policy and the way people react. They're not going to change their policy because of the size. I think we recognize this need for openness between ourselves and the customer, and within ourselves.

Standard Setting

General managers of matrix organizations are in the place where high performance standards start. We earlier identified environmental pressures for high performance through shared resources as a necessary condition for matrix organizations. But it is all too easy for organizational members to insulate themselves from these outside pressures. The general manager in a mature matrix internalizes the outside pressures and articulates them in the form of performance standards. The general manager provides direction and sets the level of effort. For the functional units, for example, this may mean beginning new development efforts into new technologies or new processes. It is only the general manager who can make the final decision on the levels to which the organization's resources in these areas will be committed. For the business units, functional commitments without a specific product focus will always be too high. For the functions, they may never be high enough. The appropriate standard is the general manager's to set. In a similar manner, when the business units' objectives are too low, or when short-term resources make a business or a product line a ripe target for internal attacks, it is the general manager who finally steps in and says "do better" or "let's give it more time" or "all right, that's enough."

This might be best illustrated with an example from Texas Instruments. One of the dimensions TI wrestles with in its sophisticated matrix is the temporal one—short-term operating performance in concert with long-term strategic planning. They call the latter an "OST" (Objectives-Strategies-Tactics) system. Referring to the action Mark Shepherd, the CEO, must take to ensure that the balance along this dimension is maintained, a senior executive explained:

> Because OST funds are supposed to be used for discretionary activities, you might think that they would be the first thing to be cut in the event of a recession. Well, we had a recession in 1970, and that spring, as it became apparent that we were going to have trouble meeting our earnings' commitments, we did begin to cut back on OST fundings. By early summer, however, it was apparent that the downturn was hitting us rather severely. The choice that top management had to make was whether or not to continue to reduce the spending on OST projects. Mark Shepherd finally drew a line, saying that he would not cut OST funds below a certain level, and would let the rest of the effect of the downturn force pressure on other elements of the P/L statement, such as operating expenses. There is no way of knowing what the right answer is in such a situation, of course, but he believed strongly that we should not mortgage our future simply for the sake of current year profitability. Subsequently, we held OST funding at about the same level in 1971 until we were sure we were out of the woods. Our funding for 1972 now exceeds the amount we were spending in 1969.[5]

Each subsystem on both sides of the matrix makes its own projections and sets specific targets for higher review. But it all comes together at the general manager's position, at the place where the overall level of aspiration for the organization is set. This duty cannot be delegated.

KEY ROLES: THE MATRIX BOSSES

In Chapter 3 we said that the matrix bosses share subordinate(s) in common with other bosses. As matrices evolve, this means that matrix bosses will find themselves on one or the other dimension in the

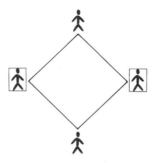

power balance. Whether the dimension is the one that is given or the one that is grown can make a significant difference for the perspective that evolves. Since one of the most typical evolutions is from a functional structure, through a project overlay, to a business–function balance, we will examine the matrix boss role for each of these two dimensions in detail. The same lessons, however, can be drawn for matrix bosses that are in charge of areas, markets, services, or clients.

The Functional Manager

One of the greatest surprises when an organization grows into a matrix comes in regard to the changing role of the functional managers. In a functional organization, managers have authority over the objectives of their function, the selection of individuals, the priorities assigned to different tasks, the assignment of their subordinates to different tasks and projects, the evaluation of progress on projects, the evaluation of subordinates' performance, and decisions about subordinates pay and promotions. They consult or take direction only from their boss in these matters, but much is self-contained in the function.

In a matrix organization none of these responsibilities remains solely the purview of the functional manager. He or she must share many of these decisions with program or business managers or other functional managers at his or her level. Many matrix structures require dual sign-offs on performance evaluations, pay and promotion decisions. Even when this is not so, consultation on these matters with others is essential for the effective functioning of the matrix and the power balance discussed above. Tasks, assignments, and priority decisions need to be shared with business managers and indeed often come from decisions made by project or business teams. Even a function's objectives are partially determined by the resource demands of projects and businesses. Functional managers in the matrix role are responding in areas in which they have traditionally been the initiators. They find themselves less in control of their areas. For example, a manufacturing manager struggled and resisted for several years the notion that many of his plant manager's goals were to be set in response to a business team's needs and that review of goal accomplishment (from a time point of view) was the business manager's and team's responsibility. He had difficulty understanding that his responsibility was to review goal accomplishment from the point of view of a functional specialist.

Thus, for functional managers a matrix organization is often experienced as a loss in status, authority, and control. It is a step down as they see it. They become less central and less powerful as parts of their previous prerogatives for initiation move from the function to the business manager. The ultimate example of this is the increased confronting of functional managers by their functional subordinates who now have membership in a business team that provides the legitimate need and the social support for such upward initiation and confrontation. For managers who have been in relative control of their domain, this is a rude awakening which can create initial hostility, and a predictable resistance to a matrix form of management.

As a matrix matures, however, functional managers adapt to these changes and can find the role not only liveable but highly challenging. Even though in matrix organization, it is the business managers who tend to control the money that buys human resources, functional managers must engage in very complex people planning. They must balance the needs of the different product lines and/or businesses in the organization, they must anticipate training needs, and they must handle union negotiations (if layoffs or promotions are involved). They must also administer support staff (supervisors, managers, secretaries, clerks) and accompanying resources (equipment, facilities, space, maintenance), many of which must be shared with the business units. To accomplish this with any kind of efficiency, functional managers must balance work loads to avoid excessive peaks and valleys in resource demands. They must do this in any organization, but in matrix, the business managers act relatively autonomously and functional managers cannot be effective by holding to some central plan prepared primarily for budget purposes. It is imperative that they know the product and business work load projections and changes well in advance; that they negotiate constantly with these managers to speed up, slow down, schedule, plan and replan the pace and amount of their activities. In other words, they must go to the business unit managers and be *proactive* if they are to manage their function well.

Some comments from several types of managers in two different matrix organizations serve to underscore this need for proactive behavior.

- Functional managers have to learn that they're losing some of their authority to product unit and they will have to take direction from the product bosses. They have to segment their work along product lines not functional lines, and they must be willing to establish communication channels with product units.

- Functional managers have to learn to become more aware of the impact of their decisions on our product–market success and become more responsive to the product organization needs which reflect the market. They have to remove their blinders and look around them while they turn the crank.

- We have to learn to serve as well as dictate; become more customer-oriented where the customer is the product line. We must realize that the function's mission is to perform the function and prove that their function is the best available. There is a burden of proof in matrix that did not exist in functional organization.

The Business Manager

As we have already pointed out, in a matrix various functional specialists are brought together in temporary (project) or permanent (business or product) groupings. These groups are led by product or business managers who have the responsibility for assuring that the efforts of functional members of the team are integrated in the interest of the project or business. In this regard they have the same responsibilities as a general executive—their objective is project accomplishment or long-term profitability of a business. However, in a matrix they do not have the same undivided authority of the general executive. People on the team do not report to them exclusively since most of the team members also report to a functional manager. Thus, as many such managers have complained, "We have all the responsibility and little of the required authority."

Top leadership in traditional organization designs has the benefit of immediate legitimacy because people understand that reporting to someone means being responsive to that someone. This is because their boss has not only formal title and status, but influences their performance evaluation, their pay, their advancement and, in the long run, their career. In a matrix, these sources of authority are shared with functional managers, thus lessening in the eyes of team members the power of the project or business managers. Project or business managers do not unilaterally decide. They manage the decision process so that differences are aired and trade-offs made in the interest of the whole. Thus, they are left with the unique job of influencing with limited formal authority. They must use their knowledge, competence, relationships, force of personality, and skills in group management to get people to do what is still necessary for project or business success.

This role of the matrix (business) boss creates both real and imagined demands for new behaviors that can be very anxiety-producing for individuals who face the job for the first time. Matrix (business) managers must rely more heavily on their personal behavior qualities, on their ability to persuade through knowledge about a program, business, or function. They must use communication and relationships to influence and move things along. Their skills in managing meetings, in bringing out points of view, and working toward concensus are taxed more than those of general managers in conventional organizations. Thus, for individuals who for the first time face these demands, the world is quite different. They can easily experience frustration, doubt, and loss of confidence as they begin to rely on new behaviors to get their job done. They begin to question their competence as they experience what in their eyes is a discrepancy between final and complete responsibility for a program and less certain means of gaining compliance from others. Some individuals learn the required new behaviors; others never do.

Not only does the actual and required change in behaviors create a problem for new matrix business managers but so also does their attitude toward the change. In our experience, individuals assigned to this role must first break through their perception of the job as impossible. Individuals who have been brought up in traditional organizations have firmly implanted in their minds the notion of hierarchy and formal authority as the sources of influence and power. They are convinced that the job cannot be done because they have never had to think through how power and influence are wielded in traditional organizations. They cling to the myth that the formal power a boss has is what gives the boss influence. This myth remains even after they themselves have developed and used other means. The myth about power and influence is often the first barrier that must be broken before the individual becomes motivated to address the real demands for new behavior.

In their relations with peers in both arms of the matrix, the posture of matrix business managers needs to be one of reason and advocacy. It is through these relations that they obtain the human resources they need to accomplish their goals. They have to expect that a number of these resources will be in short supply and that competing claims will have to be resolved. In these dialogues, business managers must stand up for their requirements without developing a fatal reputation for overstating them. They must search with their peers for imaginative ways to share scarce resources. They must reveal any developing problems quickly while there is still time for remedial action. These behaviors do not come easy to many managers conditioned in more traditional structures. An assistant project manager at TRW expressed that understanding succinctly.

> How do you maintain a relationship with the guy in the functional department so that he gives you maximum creativity and his best efforts while also getting what is required for the project? It's an influence sort of thing. You never have direct control over your resources. You have to know how the other people operate and in many cases it is a completely individual relationship with each one of the people who work on your part of the project. You work it only by having some good healthy discussions with the key people in the functional departments. The people over whom you have control in the project office also have to interface outside and be influencers in the functional departments.

Finally, in their relations with the various functional specialists represented on their team, matrix business managers must establish a balanced or intermediate orientation. They cannot be seen as biased toward one functional area. They cannot have an overly long- or short-time horizon. Their capacity to obtain a high-quality decision is dependent on an approach which seeks to integrate the views and orientations of all the various functions. If they show a bias, team members will begin to distrust their objectivity and their capacity to be fair arbiters of differences. This distrust can be the seed of a team's destruction. For many individuals, this is a very difficult task. A lifetime in a given function creates a bias imperceptible to the individual but quite obvious to others. The capacity to wear multiple hats believably and equally well creates heavy attitudinal and behavioral demands. It means having the capacity to empathize with other functions and to identify with them while at the same time maintaining a strong internal structure that guides oneself and one's role. Learning to do this is difficult and, as we shall see in the next chapter, has implications for selection, career development, and training for the role.

KEY ROLES: THE 2-BOSS MANAGERS

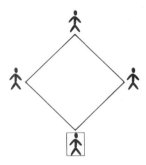

The most obvious challenge built into this matrix role is the sometimes conflicting demands of 2 bosses. For example, a representative from a manufacturing plant on a business team may know that the plant is having profitability problems and that the last thing the plant manager wants is to disrupt ongoing production activities with developmental work like making samples or experimenting with a new process. Yet, as a business team member, the plant's representative may see the importance of doing these things immediately to achieve project success. In this situation individuals in a 2-boss position experience a great deal of anxiety and stress. These come from the difficulties of weighing the conflicting interests of their function and their project team. Both have legitimate viewpoints. But which is the more important view from the perspective of the whole organization? This is neither an easy question to answer nor an easy conflict to resolve. But added to this are the questions of identification and loyalty to one's function or one's business team and the consequences of rejection or even punishment from whichever side of the matrix perceives that it has lost in a given conflict. To compound the problem, even if plant representatives on this project team decide that they need to go against what they know is in the interest of their plant, how do they communicate this back to their organization and convince the organization of the merits of their views. The same problem would exist if they were to favor their functional orientation and have to·persuade the team that sample runs will have to be delayed.

We can see from this description and our previous discussion that there are problems of dual group membership, new demands for communication, uncertainty about the kinds of commitments that can be made, uncertainties about how to influence others in one's function or team, and uncertainties created by a more generalist orientation not demanded in a conventional functional organization. There are, of course, differences in the capacity of individuals to deal with ambiguity, but all individuals new to matrix management lack some of the knowledge and the skills needed to navigate through the ambiguities and conflicts of a matrix.

The role problems of the 2-boss manager can, of course, become manageable in a mature matrix. This happens primarily because the functional and business managers learn to avoid most instances of making irreconcilable demands of their shared subordinates. This will still happen on occasion, however, even in a smoothly functioning matrix. In a familiar instance, the 2-boss manager may be directed to be in two different places at the same time. A common way of dealing with such situations is to establish the norm that 2-boss individuals are expected, and even directed, to convene a meeting between their 2 bosses to resolve any such conflict. The 2-boss manager is reprimanded only if he or she suffers such a conflict in silence.

Beyond handling such occasional problems, 2-boss managers learn in a mature matrix that their roles give them a degree of influence not usually experienced at their level in a conventional organization. They not infrequently find themselves in the balance position in a discussion with their 2 bosses over some point of conflict. If they know their facts and express their judgment on the merits of the particular issue, they often find their opinions are taken very seriously. This is the heart of training for general management. And this is exactly how the matrix is intended to work—with decisions being made at a level where the relevant information is concentrated and where time is available for a thorough airing of the options. In such a framework, a higher percentage of decisions will, in fact, be given careful attention and decided on their unique merits rather than in terms of a single orientation.

In reviewing the general characteristics of the mature matrix, we have emphasized the quality of flexibility. By looking in some detail at the key roles unique to the matrix, we have discovered where that flexibility comes from—from the individuals in those roles who have been challenged by the matrix to respond to each new situation in a fresh and flexible manner. This constant pressure for fresh thinking, for learning in the mature matrix, has, in fact, seemed to greatly increase organizations' productivity, especially at middle-management levels.

This may be fine for the organization but how about the individual? We have already seen that the matrix can cause considerable anxiety in individuals as they initially face new and demanding role expectations. Is this a problem or an opportunity? In Chapter 5 we will examine in some detail these questions. We will focus on the methods that are available to help individuals convert matrix demands into constructive opportunities.

CASE STUDY: THE TRW SYSTEMS GROUP

The origins of the TRW Systems Group go back to 1953 when Ramo-Wooldridge (RW) established a privileged relationship with the United States Air Force to perform systems engineering and forward planning on the USAF's intercontinental ballistic missile (ICBM) program. RW grew quickly by linking itself with the accelerating ICBM program, but its close relationship with the USAF prohibited it from bidding on hardware contracts. In 1958 RW joined with Thompson Products to form TRW, Inc. The part of the composite organization that was engaged in the USAF's ICBM work was retitled the TRW Systems Group. One year later, the TRW Systems Group exchanged its privileged relationship for one that allowed it to compete for hardware contracts in addition to the systems engineering activities to which it had until then been restricted.

Between 1960 and 1963, TRW Systems completed the transition from a "sheltered captive" of the Air Force to a fully independent, competitive aerospace company. During this period the number of employees increased from less than 4000 to 6000. Sales volume grew from $60 million to $125 million. Whereas in 1960, 16 contracts distributed among eight customers accounted for the sales volume, by 1963 these numbers had become 108 contracts and 42 different customers.

Although TRW Systems dropped its privileged systems engineering role, it maintained the systems management for three major ICBM programs for which it had held responsibility: the Titan missile, the Atlas missile and the Saturn system. Figure 4.1 illustrates the chart of the organization TRW Systems used in 1963 to manage these three programs (the terms project and program were used interchangeably). Each program office reported at the same level as each functional division.

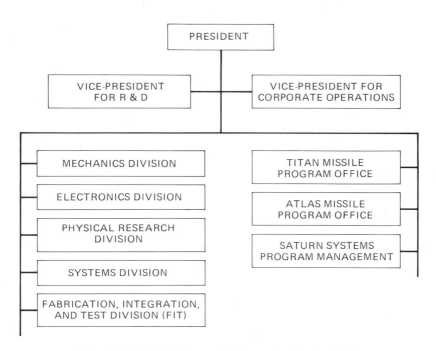

Fig. 4.1   TRW Systems matrix organization in 1963.

The Matrix Relationship

Most of TRW Systems' programs were at the state of the art of their particular technology. The Titan, Atlas, and Saturn programs and the many smaller projects they assumed called on the best technology available in rocket propulsion systems, guidance and control systems, electric power systems, telecommunications, digital systems, mechanical design, metals technology, etc. TRW Systems acquired some of that technological knowledge from government and university research and development laboratories and through a varied set of subcontractors. But they developed most of it themselves through original research and by keeping their engineers and scientists in close proximity with their respective professional bodies and sources of outside technological development. They viewed this functional capability as their primary strength.

TRW Systems' customers, however, didn't buy functional competence. They purchased a specific output that applied that knowledge via the program and project offices. TRW Systems' contracts were for projects and programs that were highly uncertain and frequently subject to changes from other components of the aerospace systems with which they interfaced. The program offices worked closely with their customers, interpreting their requirements, modifying them where necessary, and converting them to specifications they could relate to the functional resource groups.

In the case of ongoing contracts the relationships were so extensive that most customers maintained representatives physically within the TRW facilities to respond quickly to questions and crises and to monitor progress. New business, in turn, entailed considerable transaction activity with different areas of the Department of Defense and the National Aeronautic and Space Association to bid on and to influence the technical parameters on new contracts and on modifications to contracts.

The typical project office which coordinated the contract activities numbered from 30 to 40 people and was concerned with the planning, coordinating, and systems engineering of the project (see Fig. 4.2). Customer contact and responsibility for cost, schedule, and performance lay with them. The actual design and technical work was carried out by the different functional departments. A large hardware contract for a space vehicle required the integration of all of TRW Systems' capabilities to produce the systems of the subprojects. For example, the assistant program manager (APM) for planning and control was responsible for cost and schedule control, and PERT costing. In a sense APM performed the functions of a local controller and mas-

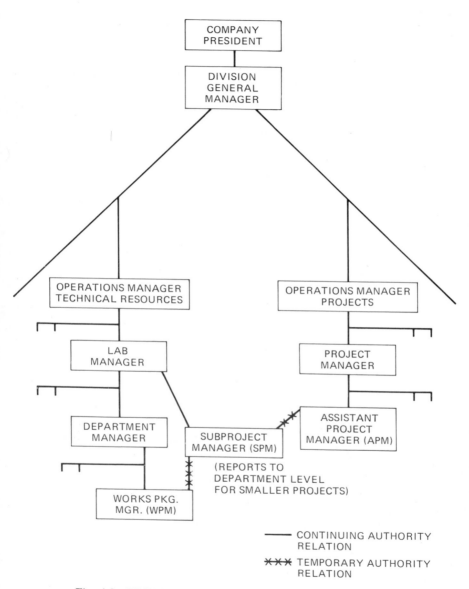

Fig. 4.2   TRW Systems group—the APM–SPM–WPM chain.

ter scheduler. The APM for systems engineering was responsible for formulating the project's systems requirements and making sure that everything was designed to fit together in the end. And the APM for product integrity was responsible for developing and implementing a reliability program for the entire project including all of its subprojects.

Subproject Management

The total project effort was divided into subprojects by the project office. A typical hardware subproject was in the $2 million to $4 million range, with an average of 50 people and a peak of over 100. The effort consisted of the analysis, design, development, and fabrication of perhaps four different assemblies comprising a subsystem. Typically, these assemblies were new designs and five to eight of each were produced over a two-year period.

Each hardware subproject was assigned to a specific functional organization. The manager of that functional organization appointed a subprogram manager (SPM) with the concurrence of the project manager. The SPM was responsible for the total subproject and was delegated management authority by both the functional management and the APM to whom the SPM reported operationally for the project. Normally the SPM reported administratively to a laboratory or department manager in the functional organization.

The SPM worked full time directing his subproject, but was not a member of the project office; he remained a member of his functional organization. He represented both the program office and functional management in his authority over the divisional people working on his subproject. He spoke for the project in such matters as scheduling personnel and facility assignments, expenditure of funds, customer requirements and design interfaces. He also, however, represented the laboratory or division in such issues as technical approach, cost-effective scheduling, and the impact of design changes. The project manager provided his evaluation of the SPM's performance to the SPM's functional manager for the SPM's salary review.

Work Packages

The SPM was responsible for proposing a "work breakdown structure" of his subproject for the project manager's approval. In this work breakdown structure, the subproject effort was successively subdivided into work packages, work units, and tasks for schedule, cost, and performance control. Job numbers were assigned at each level.

The functional engineer or supervisor receiving project direction from the SPM was called the Work Package Manager (WPM). Below the level of the WPM, the work was managed within the functional structure but the project manager maintained project control through the APM-SPM-WPM chain. The work package was generally performed entirely within one functional department.

Functional Organization

TRW Systems was organized into five operating divisions; specifically, the Space Vehicles, Electronic Systems, Systems Laboratory, Power Systems, and Systems Engineering and Integration Divisions. Each of the divisions served as a technology center which focused on the disciplines and resources necessary to practice its technology. Although each division was organized differently, they shared a similar pattern of organization.

Reporting to the division general manager were several operations managers. These operations managers were each in charge of a group of laboratories which were engaged in similar technologies. Each laboratory included a number of functional departments which were organized around technical specialties. Most divisions had a fabrication or manufacturing operations group headed by an operations manager.

Laboratories and Departments

The typical laboratory contained from 100 to 300 personnel and was engaged in anywhere from two to ten subprojects. The laboratory manager spent about half of his time reviewing the progress of these subprojects and reviewing new proposals. His main concerns also included the assignment of personnel and facilities to meet new demands on the laboratory or in anticipating impending problems. Many laboratory managers had an assistant who was in charge of the subprogram managers and responsible for monitoring the subproject work being performed in the lab. In other cases the subproject managers were responsible to department managers.

The number of departments in a laboratory might vary from two to six. A typical department could have from 25 to 100 people assigned to it, and most departments were divided into at least two sections. Few had more than five. Normally the departments were organized so that their activities were confined to a single technical specialty. The department manager was responsible for developing and maintaining the technical capabilities of his particular specialty.

While a project could be conducted entirely within one or two laboratories, work for projects which were too complex or large for one laboratory to handle were organized into project offices reporting to the division manager or an operations manager for projects.

Organization of the Matrix across Divisions

Each of the divisions at TRW Systems was a mixture of projects and functional departments and was generally organized around a fairly logical grouping of technical disciplines. Nevertheless, there was considerable cross-divisional coordination. TRW Systems approached the structural problem of which programs and which functions should go together with an "order-disorder" notion. They wanted the freedom to be able to organize in whatever way seemed best at the time—"the flexibility to be responsive"—rather than be forced into a bureaucratic logic that dictated that specific functions belonged with specific programs, etc. It was a deliberate structuring to make programs or projects go to other divisions to get support; part of a philosophy of encouraging the interdependent relationships they needed if they were to function with little or no duplication of specialties.

A working committee formed to study the TRW System organization commented on the issue of duplication:

> From the standpoint of personnel and physical resources, it is most efficient to organize by specialized groups of technologies. To stay competitive, these groups must be large enough to obtain and fully utilize expensive special equipment and highly specialized personnel. If each project had its own staff and equipment, duplication would result, resource utilization would be low, and the cost high; it might also be difficult to retain the highest caliber of technical specialists. Our customers get lowest cost and top performance in organization by specialty.
>
> For these reasons, the company has been organized into units of technical and staff specialties. As the company grows, these units grow in size, but a specialty is normally not duplicated in another organization. Each customer's needs call for a different combination of these capabilities. [Hence] a way of matching these customer needs to the appropriate TRW organizational capabilities is necessary. The use of the project office and matrix organization allows TRW Systems Group to make this fit.

Building a Project Team

The project manager of one large project described how he put together a project team.

First I look at the projects office's work load as a function of time so that no one person's work load peaks while the rest of the people in the organization have little to do. We want everyone busy with small peaks if possible.

Then I take various cuts of how we could organize the project office. If a guy is overloaded on the first cut, I adjust and so forth. It is a "real-time" thing that you assess day by day even after you have made your basic organization.

The next step I take is to look at how I am going to split up the total work along the functional side of the matrix. What departments will be needed? What will have to be subcontracted?

In doing this there are overlaps. It's then a question of defining where one subproject begins and where one ends. Another question is how big should each subproject be? Do I need two subprojects to accomplish something or do I need one? This requires a lot of thought on our part and collaboration with the functional departments. We've tried to develop some criteria over the years for defining subprojects.

One criterion is the dollars that a subproject manager is going to have to look after. Second is the number of people within the company that this man is going to direct and monitor. The third criterion is the number of technical interfaces he has. How many technical people and how many discrete technical problems does he have to work with? The fourth factor is the management interfaces involved. How many functional departments is he going to have to interface with from a management point of view? A fifth key factor would be the number of subcontracts and the nature of the procurement. Is it easy to do or is it protracted, technically detailed, demanding subcontract work? And the sixth one is the nature of the total effort. Here's where risk comes in. Does the job border on basic research where you're dealing with factors that are not yet known or is it more applied?

The Employee with 2-Bosses—The Subproject Manager (SPM)

The manager of the functional organization appointed a subproject manager with the concurrence of the project manager. The subproject manager was assigned responsibility for the total subproject activity and was delegated management authority by the functional division management and by the assistant project manager to whom he reported operationally for the project. He was accountable for performance in his functional specialty to the manager of his functional area, usually a laboratory manager. The functional manager was responsible for the performance evaluation of the subproject manager. The subproject manager thus represented both the program office and his functional area and was responsible for coordinating the work of his subproject with the engineers within the functional area. Normally each functional area was involved in work on several projects simultaneously. One manager defined the subproject manager's responsibility this way:

> The subproject manager is a prime mover in this organization, and his job is a tough one. He is the person who brings the program office's requirements and the lab's resource together to produce a subsystem. He has to deal with the pressures and needs of both sides of the matrix and is responsible for bringing a subsystem together. He has to go to the functional department managers to get engineers to work on his project, but about all he can say is "Thanks for the work you've done on my subproject." But he does have program office money as a source of power, which the functional managers need to fund their operations. The technical managers are strong people. They are not yes-men; they have their own ideas about how things ought to be done. You do not want them to be yes-men either. Otherwise you've lost the balance you need to make sure that technical performance is not sacrificed for cost and schedule expediencies which are of great importance to the program office. The functional managers also are interested in long-range applications of the work they are doing on a particular project.
>
> This often puts the subproject manager in a real bind; he gets caught between conflicting desires. It is especially difficult because it is hard for him not to identify with the program office because that's the focus of his interest. But he is paid by the lab and that is also where he must go to get his work done. If he is smart, he will identify with his subsystem and not with either the program office or the lab. He must represent the best course for his subproject, which sometimes means fighting with the program office and the departments at different times during the life of the subproject. If he reacts too much to pressures from either side, it hurts his ability to be objective about his subproject, and people will immediately sense this.

Both sides of the matrix had different forms of organizational power and those different pressures converged on the SPM. The program office had power in the form of money which the functional managers needed to fund their activities. The technical managers had the competence to make state-of-the-art projects succeed and were strong enough to ensure that technical performance was not sacrificed for the cost and schedule expediencies that underlaid the drive of the program people. To support this balance, TRW Systems insisted that the SPM remain under direct functional jurisdiction.

This kind of formalization was necessary because it was too easy to get completely committed to the project's shorter-term goals. One SPM described this tendency.

One of the dangers of being in this job is that you identify too much with the project office. You can't become so identified with a project that you lose sight that the guy who is in trouble is in your department. It's possible to be so program-oriented that you're throwing stones at your own guys. You're the front-line representative of the program office but you're still getting paid by the lab manager. . .there is no question in my mind that he's my boss.

Referring to the pressures of being in the middle, one ex-SPM described the SPM as "the one person in the functional area who ensures that costs and schedules are maintained for a part of the project." Nevertheless, he was still able to state, "The SPM's job is probably one of the most uncomfortable yet rewarding positions a man can have."

Managing the Matrix

Jim Dunlap was the Director of Industrial Relations during the 1960s, the formative years for TRW Systems' matrix organization. He described the complexities of managing in a matrix where people have 2-bosses.

The decisions of priority on where a man should spend his time are made by the president because he is the only common boss. But, of course, you try to get them to resolve it at a lower level. You just have to learn to live with ambiguity. It's not a structured situation. It just can't be.

You have to understand the needs of Systems to understand why we need the matrix organization. There are some good reasons why we use a matrix. Because R&D-type programs are finite programs—you create them, they live and then they die—they have to die or overhead is out of line. Also, there are several stages in any project. You don't necessarily need the same people on the project all the time. In fact, you waste the creative people if they work until the end finishing it up. The matrix is flexible. We can shift creative people around and bring in the people who are needed at various stages in the project. The creative people in the functions are professionals and are leaders in their technical disciplines. So the functional relationship helps them to continue to improve their professional expertise. Also, there is a responsiveness to all kinds of crises that come up. You sometimes have 30 days to answer a proposal—so you can put together a team with guys from everywhere. We're used to temporary systems; that's the way we live.

Often an engineer will work on two or three projects at a time and he just emphasizes one more than others. He's part of two systems at the same time.

The key word in the matrix organization is interdependency. Matrix means multiple interdependencies. We're continually setting up temporary systems. For example, we set up a project manager for the *Saturn* project with 20 people under him. Then he would call on people in systems engineering to get things started on the project. Next he might call in people from the Electronics Division, and after they finish their work the project would go to FIT [Fabrication, Integration, and Testing] where it would be manufactured. So what's involved is a lot of people coming in and then leaving the project.

There is a large gap between authority and responsibility and we plan it that way. We give a man more responsibility than he has authority and the only way he can do his job is to collaborate with other people. The effect is that the system is flexible and adaptive, but it's hard to live with. An example of this is that the project manager has no authority over people working on the project from the functional areas. He can't decide on their pay, promotion, or even how much time they'll spend on his project as opposed to some other project. He has to work with the functional heads on these problems. We purposely set up this imbalance between authority and responsibility. We design a situation so that it's ambiguous. That way people have to collaborate and be flexible. You just can't rely on bureaucracy or power to solve your problems.

NOTES

1. William C. Goggin, 1974. How the multidimensional structure works at Dow Corning. *Harvard Business Review,* Jan./Feb.
2. See Texas Instruments Incorporated, International Case Clearing House, 9–172–054, Boston, Mass. 02163.
3. In an address at the Harvard Business School, Boston, Mass. March 12, 1975.
4. Texas Instruments Incorporated, op. cit., p. 5.
5. Texas Instruments Incorporated, op. cit., p. 17.

5
MATRIX AND THE INDIVIDUAL[1]

In the preceeding chapters we have described the concept of a matrix—what it is, how it develops, and how it works. The discussion has been mainly of new structures and new roles. There has been substantial emphasis on the need to design the organization so that people will behave in the manner intended by the structure. In this chapter we want to explore more fully the demands that a matrix organization places on people. It should be clear from the previous discussion that a matrix demands new behavior, attitudes, skills, and knowledge. This fact, in turn, has substantial implications for selection of people into the organization, their development, and their career progressions. We will explore these demands and how an organization might help individuals and groups cope with them. The success of a matrix depends on the capacity of the organization to help people learn how to functions in new ways.

But the demands of a matrix organization also create opportunities for people as well as the organization. People can grow and develop in ways and at rates not normally possible in more traditional organizations. There are opportunities for growth in knowledge, skill, interpersonal competence, and influence. There are opportunities for the organization to reap greater motivation and commitment from organizational members. A matrix is a high-tension system which places greater demands on people, but which offers a bonus in newfound opportunities. In the discussion which follows, we will be exploring the demands and the opportunities for people.

MANAGING CONFLICT

It should be quite clear from the discussion of new-role demands that a matrix increases the amount and the pattern of required contact between individuals. Peers in different functions have more contacts because of interdependencies around projects. Communication must also increase within functions as individuals at the 2-boss position obtain information from their functional organizations and attempt to influence them. The interdependencies of a matrix simply require increased communication.

But this increased communication creates a new set of problems. More differences are surfaced and they have to be dealt with. It is not that differences do not exist in functional organization. They do. But, in the functional organization, roles are structured so that individuals can usually resolve conflicting demands by talking to their own functional boss. In a matrix these differences are resolved with people from different functions who often have different attitudes and orientations. They are resolved without a common boss readily available to arbitrate differences. Top managers quickly become overloaded when too many decisions are passed to them and the matrix ceases to work.

The assumption in a matrix is that this conflict can be healthy and that higher quality solutions will develop if people with different expertise and orientations relating to a given task get together to thrash out their differences. There is a demand for confronting and problem solving. In this approach to conflict, management differences are valued and people express their views even when they know that others may disagree. It is assumed that everyone is working toward common goals and that each is concerned with arriving at the best solution to the problem. Given the constraints of the situation, if sufficient information is brought out in the discussion, the best possible solution will emerge. The confrontation–problem-solving process is dependent on all persons' airing their agendas for why they favor a given course of action so that these can become part of the public domain and be examined. It also depends on individuals' willingness to abandon a position and take up another in the discussion, so that they can examine their original position from a different perspective and be less emotionally committed to their starting position.

In a confrontation–problem-solving discussion, people air sharp differences but without hostility or destructiveness. There is substantial fluidity in the positions people take with individuals shifting positions and presenting new alternatives as the discussion progresses. There is also a substantial amount of questions aimed at obtaining more information or gaining clarification of statements and positions. There is a minimum of entrenched advocacy and of quick and unilateral evaluations. Premature evaluations tend to reduce openness and a willingness to air differences. Unfortunately, many conventional managerial groups tend to be highly evaluative in their communications. Thus, the need for a confronting–problem-solving process of conflict management implies a need for changes in interpersonal competence and group process. This, as we shall see later, requires substantial training and development efforts aimed at changing both organization culture and individual behavior.

The confrontation–problem-solving approach to conflict can be contrasted with other approaches also used in dealing with conflict. When differences exist, it is possible for individuals to make believe that these differences do not exist and to act in a way that maintains cordial personal relationships but does not deal with underlying differences. This is called *smoothing* and is often used to prevent emotional confrontations that might get out of control. Parties in a conflict could also *bargain* or *compromise* to resolve a difference; each gives a little but neither gets everything. The compromise solution may meet individual needs partially, but it may not be the best solution for the organization. It is also possible for individuals to *withdraw* from a conflict situation, but this is less likely to occur in a matrix where frequent contact is demanded. Finally, differences can be dealt with by one party *forcing* another to submit to superior power based on knowledge or authority. The message sent is, "You comply or there will be sanctions." In an effective matrix organization, confrontation is the primary mode of conflict management, and forcing or compromise the second most frequently used mode. Forcing, while reducing trust, does at least assure that some action is taken. Smoothing and withdrawal occur less frequently. People operating in a matrix need to understand this and have the skills and the interpersonal competence to manage the conflicts that are surfaced.

COLLABORATION AND TRUST

Despite greater opportunities for conflict, a matrix organization demands high levels of collaboration. Top leadership of the matrix must work collaboratively in making decisions on strategies and resource allocations. They must jointly agree on their expectations of various matrix managers and together evaluate the managers' effectiveness. While this is true in every organization, it is especially true in a matrix because matrix managers make more operational trade-off decisions without any further reference to top leadership. Different and inconsistent signals will be more costly in a matrix.

Similarly, matrix and 2-boss managers will have to work collaboratively to integrate their different viewpoints. This places greater demands on individuals to share information, to work together on tasks such as planning, to consult others before making decisions, to think about the whole and not only their function. Often this notion of teamwork is interpreted as simply the pooling of individual plans and activities. Some individuals do not recognize that a matrix is adopted to deal more adequately with interdependence. Here is an example of a group whose view of working together is limited to sharing information. In developing their yearly plan for a major new venture, each individual on one team develops his or her own functional plan and gives it to the venture manager for consolidation into an overall plan. All the interactions are one-on-one between the venture manager and the team members. There is no recognition that developing a plan is an interdependent task in which functional plans, once developed individually, must be shared so that differences can be reconciled and joint agreements reached. The task of planning requires high levels of collaboration. While business teams do not always need to meet as a group, a matrix induces more of this than do other organizational forms.

Collaborative behavior is also needed between groups. A top-management group needs to meet periodically with various business or functional groups to review their progress. This process must go beyond the normal review of numbers and objectives. The review process must allow a dialogue between the top group and the team so that both groups can go away with an understanding of each other's viewpoints and problems. If this understanding does not develop, inevitable surprises do. Deadlines are missed, budgets are overrun, and learning curves on new programs are longer than expected. A matrix exists to manage uncertainty. Only a collaborative process and open dialogue can enable a top-management group to get a "feel" for the uncertainties a lower level group might be facing. This "feel" allows the group to better assess the tasks and probabilities of success inherent in a business or program, and to respond to a team's needs for guidance and direction appropriately. The lower level team needs less checking than it needs help in making decisions that are in line with the larger objectives of the organization. These are problem-solving discussions and helping–coaching sessions. These are elements of good management, in general, but the demand for these types of behavior is far greater in a matrix than in a unity of command organization operating in a more certain environment.

To achieve collaboration, the needed ingredient is trust. Individuals and groups must learn to rely on each other and to accept each other's judgments when these are based on unique competence and knowledge. Without trust, the organization quickly reverts back to a reliance on chain-of-command authority. Individuals share less, make more unilateral decisions, use forcing as a way of resolving conflict, and see greater differences between themselves and those in other functions. Differences in attitude and orientations are magnified and stereotypes develop.

To develop and maintain trust, individuals in a matrix must be prepared to take personal risks in sharing information and revealing their own views, attitudes, and feelings. For example, if matrix members reveal the uncertainties they feel due to conflicting demands between function and project or business, others will tend to open up more. On the other hand, if individuals hold back information in an attempt to protect their position, distrust will tend to develop.

In an effective matrix, people continuously work at taking risks and communicating openly. Not to do so can start a negative cycle. Open relationships cannot be taken for granted. This does not make a matrix a "comfortable" organization in which to work. Individuals must push themselves to reveal more about themselves and their function's position than is common in unity of command organizations. The discomfort comes from the fact that once they have revealed themselves, they perceive themselves to be less in control of the final decision reached and of their relationships with others. Did what I said hurt somebody's feelings? Have I revealed so much about the problems in my group that we will be taken advantage of? While this is not true if others are equally open and trusting, the feeling of being out on a limb and exposed is sufficiently anxiety-producing that it provides a continual barrier that must be overcome. It takes more emotional energy, at least in the beginning, to work in a matrix.

As a final point, it must not be forgotten that the matrix demands sophisticated problem-solving skills from those involved. Analytic and cognitive skills required are of the same order, even if they do not convey as large a consequence, as that required of top management. They must scrutinize quantitative and qualitative data. They must be prepared to use relevant techniques without exaggerating their utility. They must be aware of broad ethical, as well as of purely economic, considerations. They must weigh short- and long-term considerations. They must accurately anticipate responses from numerous quarters. They must make judgments of considerable consequence in the face of many uncertainties. The development of these skills constitutes a significant demand. Of course, such problem-solving skills are needed in every organization, but the essence of the matrix is that more people, in lower levels, will be involved in this decision making.

DEVELOPMENT METHODS

The previous sections outlined the major changes that matrix implies for individuals and groups. New behavior, new attitudes, new interpersonal skills, and new knowledge are demanded of people. Greater emotional energy is required because people must be more open, take more risks, work at developing trust and trusting others. As a functional manager once said about working in a matrix, "You are asking us to commit an unnatural act." He was experiencing at the time the gap between required behaviors and attitudes and his own and others' actual behavior. Does this mean that a matrix organization is an impossible dream? Is it an approach that makes sense theoretically, but is impossible to implement practically?

Experience with matrix suggests that its implementation is not an impossible dream. Individuals can and do learn how to behave in new ways. What is also clear is that they need help in doing it, and that it is particularly difficult in the early phases of a matrix before critical mass in the change process has been achieved. It is in this early phase when individuals feel that they have been asked to change their ways of doing things, but it seems as if no one else has. Unless individuals and groups receive support and help in adjusting to a matrix in this early phase, the matrix probably will fail.

In the case of one attempt to create a multinational matrix, the roles of worldwide business managers were created to integrate planning and decision making between the foreign and domestic parts of the company, but the individuals placed in the new role were never counseled and coached about what the new roles demanded. Thus, they interpreted the job in accordance with their personal inclinations. Only one business manager was successful and this was because his natural style was to work collaboratively with others, to create group situations in which he could facilitate the decision-making process, and to help link and bring people together. Another business manager followed his natural inclination to take charge and promptly walked into an overseas subsidiary and told them what he wanted done. The flack that resulted from this behavior created suspicion in the subsidiary and permanently damaged the person's ability to function as a business matrix boss. In the same organization, many other unresolved conflicts developed between the area managers and the new business managers. There was enormous frustration because no time had been taken to bring area and business managers together to help them sort out their roles. The ambiguity of roles resulted in emotional costs for individuals, and lost effectiveness for the organization.

It is not sufficient to merely tell people that they will be shifting from being a conventional line manager to a manager in a matrix. If they are to be effective, they must quickly build effective working relationships with the others in their matrix. It is too risky to let chance events in their contact form the character and process of the group. If left to chance alone, the group or individuals in it may develop patterns which destroy trust and reduce the potential effectiveness of their relationship. Furthermore, individuals left to their own devices do not benefit from what is already known about issues that are generally faced in the process of matrix development. A planned process of both individual and group development can be used. We call this team banding and disbanding.

TEAM BANDING AND DISBANDING

A planned startup for a matrix business group involves members meeting anywhere from several hours to a day to reach agreement on their objectives and on how they will function as a group. A team-building meeting aimed at helping a group get started might include the following agenda items:

- Group members begin a dialogue by talking about their expectations and concerns for the team and the project. This inevitably causes differences to surface which need to be recorded and worked through.

- A discussion aimed at developing agreement on the objectives of the group.

- A discussion leading to agreement on how frequently the team will meet, where it will meet, and members' expectations about attending.

- A discussion about leadership in the team—the role of the chairperson, others' responsibility for initiative in meetings and outside meetings, and the extent to which the group expects the leader to push the group to a decision.

- Roles and responsibilities of group members are discussed, with the aim of recognizing ambiguities and overlaps, not of entirely eliminating them. There are a number of techniques for clarifying roles which have been used. For example, the group can list all major decisions that have to be made and chart who has responsibility for a decision, who participates in it, who is consulted, and who must approve. In this way team members gain clarity about how their functional responsibilities relate to those of others.

- The group discusses how decisions will be made. Will unanimity be required? Is consensus sufficient and what does it mean? Will a voting procedure be used? Or will the leader make the final decisions? How will the next level up be involved?

- Ground rules for communication and conflict resolution are developed. For example, does the group want to foster an open airing of differences? What are members to do when they disagree strongly with the team and haven't been able to influence it? How are members to handle information about what the team is doing?

- Understanding is developed about the responsibilities of team members in relating back to their departments what is happening on the team.

- Any interpersonal problems are aired so that they do not block team functioning.

The first meeting of the team should be to start the team-banding process. A professional staff person with experience in team banding can help team members identify what they need to talk about. The staff person may interview individuals before the first meeting to identify their concerns about the task, about each other, or about management's attitude. This is done so that all relevant data are sure to be dealt with in the meeting. The important thing is for team members to spend a concentrated period in discussing the issues listed above, in order to develop common expectations and to prevent surprises which often lead to distrust. Discussion of potentially sensitive issues also creates a climate of openness which allows the group to deal more effectively with future problems.

The discussion of team banding is in no way meant to imply that a matrix requires people to work in groups at all times and on all problems. One of the mistakes made by individuals in the early development of a matrix is to assume that all decisions and all activities associated with a project or business must involve the whole group. There are many decisions that involve only one line of command or require communication between only two or three team members. A team-building process can help clarify which activities should be group activities and which should be handled on a one-on-one basis. In Chapter 6 we discuss how to cope with "groupitis." A team can also receive help from a process consultant who sits in on their early meetings, observes, helps them examine how they work as a group, and who might facilitate the confronting of a problem.

Processing of meetings and team banding are methods whereby individuals learn new behaviors, skills, and attitudes by dealing with real-time problems. These are not classroom training methods. Increasingly, companies that are installing a matrix are finding these methods useful in helping individuals to manage effectively.

The changing nature of work in matrix organizations means that teams must continually be disbanded. This is as much an inevitable part of organizational life as team building, but it tends to be given very little systematic attention. It is all too often assumed that the breakup of an established team can be taken for granted without any need for managerial attention. This is true even though organizational study has repeatedly indicated that one of the important sources of resistance to change is the understandable reluctance people feel about giving up long-established work relationships. The authors know of no firm using a systematic educational approach to disbanding, but we are aware of the practice of many skilled managers to deliberately help a team find an appropriate way to terminate its activities. Breaking up a successful and satisfying work group is bound to generate feelings of loss—sometimes very strong ones. These feelings need to be acknowledged and accepted. There is a need for some thoughtfully arranged ceremony that provides recognition for the team and helps all concerned work through the closure process. The handling of this process with grace is an important test of leadership.

MANAGER DEVELOPMENT

The team-banding methods described above are powerful in their impact on individual and group behavior, particularly in changing attitudes about collaboration. Yet there is knowledge and there are skills needed by *individuals* in a matrix which are not obtained in the team-banding process. Manager and training development should augment learning from team building. One of the conspicuous educational needs that is too often forgotten is an understanding of the nature of the matrix. Many of the points covered in this book can be linked to the specific *local* reasons a matrix is being adopted.

The first time an engineer, scientist, or salesperson has been involved in analyzing markets, assessing technology, or developing a business plan may occur when he or she becomes a member of a business team. Greater demand is placed on individuals for analyses of business problems, problem solving, and planning. There are numerous programs which are aimed at improving the problem-solving capability of individuals, particularly the capability for solving business problems. The interdependence of many activities in the accomplishment of a single task can require sophisticated planning techniques. Methods such as CPM and PERT, market research and analysis, financial analysis and budgeting are commonly used and can be taught by classroom methods.

It is possible to design a training program for matrix managers that incorporates individual skill training, experiential training, and a team-building startup. Such a program might include the following elements:

- Knowledge input about matrix organizations and information about why their organization is adopting matrix. This would include top management's philosophy.

- Lecture, discussion, and exercises about effective communication and group process.

- Lecture and exercises on concepts and techniques relevant to the kind of business problem solving expected.

- A simulation in which individuals are randomly placed in groups and given a business task. Each member is given a role to play and each experiences directly the problem of making and implementing decisions. Each examines the experience with the help of a trainer and learns from it.

- The actual teams are brought together to work on a number of exercises to create a low-risk setting for self-examination and learning.

- A team-bulding meeting is conducted, with a professional, as a process consultant, to go over the important startup questions.

COMPATIBLE AND INCOMPATIBLE PERSONALITIES

Despite all development efforts, some individuals cannot adapt to the behavioral and attitudinal requirements of the matrix. People are different, and their capacity to adapt is limited by their personalities and the cumulative effect of their work experience. It is inevitable, therefore, that a matrix organization creates some "dinosaurs"—individuals who may have been competent in a traditional organization but do not want to or cannot function in a matrix.

I think you find more frustration with the matrix in the manufacturing and scheduling operations where people tend to be oriented to more traditional organizations. For example, I can think of one guy who is a real "heavy" in manufacturing and he is very unhappy here because he's been used to calling the shots and having people jump, or, if there is a problem, going up the ladder to get a definite answer.

We have already cited the example of the business manager in a multinational matrix who marched into a subsidiary company and gave orders about what was to be done. Developmental experience of that individual's background indicates that a careful job of selection could probably have averted his ultimate failure. He had spent all of his career as line manager, never experiencing the problems of influence from a staff position. Furthermore, it was known to some in the company that he had been a very authoritarian and controlling general manager in his previous position. His past experience and information about his behavior could have been used to predict that he would have difficulty in a key matrix assignment.

As we have seen in our earlier discussion of the matrix boss, the position requires someone who:

- has knowledge of all functional specialties, particularly those that are most complex and uncertain. These are the ones in which judgments, rather than analysis, guides decisions.
- must be able to assess the judgment of the functional specialists and know how to challenge their positions if they are thought to be biased;
- is motivated to work collaboratively and has the skills to do it;
- is unbiased in orientation toward other functions and can work and maintain a balanced orientation;
- uses personality and expertise as a source of influence even when formal power is available;
- involves others in decision making rather than making sole decisions;
- is *not* dogmatic and impatient with the participative problem-solving process.

Our earlier discussion also indicated that there are certain characteristics all 2-boss managers and their superiors must share:

- inclination toward collaboration and skills to go with it;
- higher levels of interpersonal competence than are normally required;
- the capacity to work with dual pressures without becoming totally captured by either;
- the capacity and skills to engage in problem solving in groups;
- the capacity to develop and maintain a broad organizational perspective.

Not all individuals have these capacities. As we have seen, bosses who share subordinates must support a matrix by accepting incomplete responsibility and authority and by supporting the values of doing so through their own behavior. There are numerous examples of managers who have been effective in a traditional organization but who become increasingly critical of a matrix and the ideas of shared responsibility. In the early phases of matrix management, these individuals can block the development of required behavior within the matrix. As the matrix gains momentum, they strike a discordant note and become isolated. A sales service manager in one matrix was ultimately asked to leave the organization for these reasons despite a prior record in building a successful service organization. Not surprisingly, he went into business for himself. He had become a "dinosaur" in an organization where individual decision making was less important than collaborative problem solving.

As a matrix develops, individuals who are not at all comfortable with it begin to select themselves out. They simply do not believe that such an organization can get anything done. This process of self-selection should be encouraged. Many of the training and development activities described earlier can help the self-selection process. In going through them, people can be helped to make determinations about the fit between themselves and the matrix. Furthermore, self-selection can be an important screening mechanism when hiring new people. As new people are recruited into the organization, they should be told explicitly about the matrix—its demands and opportunities. If the person is to have 2-bosses, share subordinates, or be an integrator, he or she should talk with people in equivalent positions in the matrix.

There are a variety of ways in which compatible personality and intellectural characteristics we have described can be measured. Administering paper-and-pencil personality inventories is one way. In-depth interviews are another. In the past ten years assessment centers have increasingly been used to predict managerial potential in corporations. This same technology can be applied to the selection of individuals for matrix organizations.

Performance appraisal records are also potentially very useful in making placement decisions about what kind of person should occupy key roles in a matrix. This is particularly true when the appraisal system includes ratings of behavior which indicate how individuals go about accomplishing their tasks. Behavior which is important for effectiveness in a matrix can be specified, and information about the extent to which any individual exhibits it can help decide the fit of that individual with various roles.

HUMAN RESOURCE PLANNING

A matrix creates demands for improved planning of human resources. The requirement for a large number of matrix bosses means that the organization would be wise to provide for job experience in and between the two organizing dimensions *early* in the careers of those who show the most potential. Unless this is done, it is often too late or practically impossible to provide the necessary dual training late in an individual's career. The chances of success in other functions, businesses or markets, are smaller because attitudes and behavioral patterns are set. Furthermore, individuals who have already moved up in their initial functions are typically paid a salary that is out of line with the salary range of a lower level job that might be required to gain experience in another function. A system for early identification of potential matrix managers and a plan for their career progression is essential.

Though matrix bosses pose the most pressing problem in human resource planning in the matrix, the demand for effective 2-boss managers also requires that the organization plan more carefully. While traditional organizations may be able to sustain themselves without such planning, survival of a matrix organization may rest on it, particularly in the early years. The organization must examine all of the critical matrix roles and assess which of the current incumbents are behaviorally suited to these roles. For those that are not, realistic development goals must be created jointly with these individuals and potential replacements should be identified in case these individuals need to be reassigned. All of this should be done in an open way so that individuals are aware of what is expected of them and what alternatives are available to them.

If an organization moving to a matrix is part of a larger corporation, the human resource plan should specify at what levels people will be brought in from other divisions. Because a matrix requires attitudes and behavior not likely to be developed in line-and-staff organizations, it will probably be wise for the division with a matrix to bring in people at lower levels and grow them internally. In this way the possibility of a misfit between key manager and his or her other matrix role is reduced.

Another problem faced by matrix organizations is the shift in the value ascribed to jobs (job-evaluation rating). One of the biggest early adjustments in a matrix is acceptance by ex-line managers that business matrix bosses have jobs with ratings and salaries that are equal to, or greater than, jobs which have been seen traditionally as more important. In one organization a matrix did not work as effectively as it could because business matrix jobs were not evaluated as highly as the functional matrix bosses, thereby making it impossible for the organization to attract experienced and high potential people into the role. The quality of personnel along each dimension must be balanced.

COSTS AND BENEFITS

We have touched on only some of the management and organization development options available to facilitate evolution of a matrix. There are numerous other technologies and training programs that can be used creatively. The key point is that the demands the matrix places on individuals can be addressed, indeed must be addressed, with a variety of programs that enhance the capacity of people to function in *new* ways.

These programs, however, are not free. They require both large amounts of time away of the involved managers and substantial professional training resources. An organization of even moderate size might require one or more full-time training professionals. For example, in a unit with sales of $30 million, two professional organizational consultants were assigned for the first two years. One continued for several more years as the unit moved from Phases III to IV.

Organization and management costs are real, but so are the potential payoffs. Providing the training that gets a matrix plan off to a good start can make all the difference. Teams may be merely communication devices or fully integrated decision-making groups that take on profit responsibility. Similarly, people can adapt marginally to a new way of functioning or they can develop significantly as individuals and managers. It is our experience that one of the biggest benefits of the matrix is the developmental opportunities that it provides for members of an organization. Understanding of other functions is enhanced in ways that only job transfer has been able to achieve in traditional organizations. Skills relating to other people and working in groups are sharpened. Skills in planning, analyses of business problems, and decision making are developed. For many individuals, becoming a member on a project or business team may be the first time their horizons have been signficantly broadened.

A matrix organization allows more individuals the opportunity to develop from technical or functional specialists into general managers. Thus, the organization has an increased number of people from which they can choose their leaders. It is not unusual for people to report that their membership in the matrix provided the "best management development experience they had had in the company." The demands for knowledge of different functions, businesses or markets, and the involvement in decisions, challenge people and create high levels of commitment and motivation.

Though the matrix organization speeds the development of individuals from specialists to generalists, this can create new problems. Once individuals become business matrix bosses, for example, they will often find it difficult to move back into functional management roles. Thus, the organization is faced with the problem of creating opportunities for an increasing number of generalists. Depending upon growth rates, many of these may be accommodated in the matrix organization by a career path which leads from smaller to bigger projects and businesses. Other individuals may become potential general managers for other divisions of a large company or may find careers in other organizations. Indeed, it is possible that adoption of a matrix will create an oversupply of potential which ultimately will lead to people leaving for promotions and better opportunities.

An organization with a matrix must be alert to the changes that will occur in the personal growth and career goals of its members and accommodate these changes with adjustments in career paths, job evaluation systems, and in human resource flows.

THE HUMAN IMPLICATIONS OF A MATRIX

We started this chapter by referring to a matrix as a high-tension system. The meaning of that statement should be clearer now. A matrix demands new skills, attitudes, and behavior, but in doing so it also creates opportunities for personal growth and development unmatched in traditional organization designs. However, these opportunities do not accrue to everyone. For the most part, they do not accrue to the rank and file beneath the 2-boss manager, except in so far as the more flexible and open climate permeates down through the organization. Also, only those who have the capacity to grow and develop obtain the developmental bonus of a matrix. For some, as we have seen, the matrix may mean loss of further opportunities for promotion, demotion, or even departure from the organization. Thus, changing to a matrix does have some human costs associated with it. These can be minimized by the developmental programs we discussed and by an open dialogue with individuals about where they stand.

But what of the individuals that learn to live in the matrix? The higher levels of ambiguity and conflict inherent in matrix life must be balanced against the developmental opportunities. Each individual must exert more personal energy to define his or her role, negotiate conflicts, and make personal decisions based on judgment about the appropriate coordination and direction of an activity. He or she must take more personal responsibility and can rely less on the organization to dictate what to do. This creates more stress, but it also creates more freedom and power for individuals. Individuals in the matrix can have more influence because they may have 2-bosses who may not agree. By making one's own decisions about the appropriate directions and bringing one's 2-bosses together, the individual has the opportunity to negotiate an agreement that reflects his or her own position. The opportunity exists for upward influence.

The same thing is true in the context of teams. A business team that has agreed on a direction based on valid data has an opportunity to influence the top of the organization to an extent unprecedented in a functional organization. Whether such increased influence and freedom is a net gain when weighed against increased stress probably depends on the individual and his or her values. A matrix presents the trade-offs found in other walks of life; more responsibility and freedom is accompanied by more stress and higher demands for energy.

For the organization, a matrix means that more and more responsibility is being taken at lower levels. As a result, not only are better decisions made but more involvement leads to more commitment. We would expect, therefore, that a matrix would have available to it more potential human energy. This would be felt by the organization in terms of higher individual productivity and greater organizational effectiveness. The organization should be more responsive to its environment and experience fewer surprises. Thus, despite some fears that the ambiguity of the matrix means less control, the organization should experience higher levels of control though individuals may be controlled less.

A matrix means considerable change for individuals, and the sum of these individual changes means cultural change. We have attempted to give some feel for the magnitude of this change and to point to some means for managing it. Attention to selection, management, and organization development interventions, and increased efforts in human resource planning and career development can make the difference between success and failure for individuals in the matrix, and indeed for the matrix organization as a whole.

In the case that follows we will draw again on the experience of TRW Systems to portray the more personal responses of managers as they adapt to a matrix. We will quote at some length a mix of individuals who are learning to cope with the various roles of the matrix diamond. TRW Systems not only was one of the pioneer firms in adopting a matrix but also developed a number of programs to assist people in the transition process. Their experience with these methods is also reported.

CASE STUDY: MATRIX AND
THE INDIVIDUAL AT TRW SYSTEMS

How people have responded to matrix management at TRW has varied. The direct quotations below present a range of these reactions and also highlight some of the problems of accommodation that face individuals in the key matrix roles. TRW Systems has organized a number of educational experiences to assist people in adjusting to the matrix and achieving effective working relations. Examples of those educational programs are also provided.

From 2-Boss Managers (Subproject Managers)

The 2-boss job is probably one of the most uncomfortable yet rewarding positions a person can have. I think it needs strengthening the most, with more support staff. Let's face it, that's where the interfacing takes place between the line and the project office and in large programs that's where most of the technical direction takes place.

Although the project office and lab can put pressure on a subproject manager, both organizations know that if you put too much pressure on that point, it will break down. . .especially because it has so much stress on it already.

A subproject manager has to be competent technically to communicate with people about detailed technical problems. I think I had this respect because the people in the laboratory didn't want me to transfer to the project side of the house, and tried to talk me into staying. It's hard to explain, but I felt that when I was a subproject manager my own position as a member of the laboratory was unchallenged because people accepted me and realized I was a capable guy and a pretty good engineer. I guess it was easier for me to look after the best interests of the program office, because I didn't have to worry about my standing as a member of the laboratory.

Another subproject manager described the role he played as the intermediary between the project office and the functional area:

I have asked our people to treat the project office like a customer, that is, honest but discrete. I've encouraged contact between our working guys and the project office for informations purposes. All other things and the technical direction come through me.

I have conflicts with other subproject managers. It's fertile for problems when two projects are coming down the same assembly line. We're all TRW. . .what the hell. . .it's a question of figuring out who has the worst problem and helping each other. Sometimes I have problems working in another lab, and I have to make sure that we're getting the internal management that we need without being there eight hours a day.

Another subproject manager felt differently about the subproject's conflicting loyalties:

The question of divided loyalties doesn't really come up very often. I feel responsibility for both the program office and the functional area. I won't carry people for the functional department for free on a project, for example. But, on the other hand, I won't push people for an early completion just to feel safe—especially when it means these people are going to be sitting around after it's done. I've found that if I'm objective with both sides, and focus on the subprogram's needs, I'm not squeezed. A good part of this is because my lab manager says, "Your charter is to look out for your subprogram, period."

I'm sure there are several labs that have their subproject managers more functionally oriented than we are and it's not because the subproject manager wants to be. It's because the lab manager wants him to be. If the lab manager wants a subproject manager to be functionally oriented first and foremost, you can say anything you want and write all the reports you want, they will be functionally oriented.

A fourth SPM described the frustrations and satisfactions of the job.

I get my greatest satisfaction when we're able to carry off the plan with a minimum of changes. My frustrations, they're legion: organization, personal relationships, misinformation on overruns, adapting to constant changes in design plans, schedule, etc. But it has to be that way because that's the nature of the job. I enjoy this kind of excitement and going to meetings. It's part of the romance of the job.

There are a lot of binds that an SPM can find himself in. The most typical one is when there is a difference of technical opinion between the project office and the functional department manager.

From Functional (Engineering) Matrix Managers

One of the ways in which being an engineer here is different from other places is that you have technical responsibility and, in effect, no authority. We've had problems on the shutoff valve. They were technical problems which I, as a development engineer, felt needed correction. But the project office thought, on the other hand, that it was satisfactory the way it was and that no changes were required. It's not so much the dollar that the project office is concerned with as schedules. . . .

It's very difficult when you feel you have the responsibility, knowledge, and experience, but not the funds to authorize more work. You have to convince other people who are not familiar or experienced that it's necessary. Unfortunately, with valves, it's hard to justify them until you've had a series of failures and have actually stopped delivery of an engine. When this happens they sometimes look at you and say, "Why didn't you push harder?" It's happened here but they've always been man enough to accept the responsibility.

A manager of another development department explained that the orientation of engineers was to produce a good technical product (sometimes at the cost of lengthening the schedule and exceeding the budget), while the project office "pulled too hard the other way." He commented:

The engineers don't really understand the project implications nor are they interested in controlling the project. The project people don't really understand the technical implications of what they are doing, and often don't present technical things well enough to the customer nor screen the customer's unreasonable demands from getting into the work. So natural conflict begins to arise.

Another development department manager explained some of the difficult issues that arose:

There have been times when nobody has known clearly what to do, and where the pressures have been intense: budgets exceeded, schedules slipped, problems almost unbearable, personal work loads high, and tempers short. You literally get to the point where you may have some nervous breakdowns. The force of a few personalities becomes very important at that point. The same people are not quite so important when things are running smoothly.

The matrix requires you to be aware of the individual you are dealing with in judging the way they present their case. . . . We have had problems when personalities are not well balanced. It's a bad situation when one personality is much stronger than the other. The stronger begins to dictate, and the balance between line and project office is lost. We had a very strong development engineer in one area and his counterpart on the project side was weak. The development engineer was in control. No doubt about it. The project office in that situation was providing service, keeping the budget and documents straight, and everybody worked for this development engineer. At exactly the same time there was another area where the project engineer was a very strong personality and the development engineer was not. The project was in absolute control there because it controls the funds to begin with.

From Program (Project) Matrix Managers

I have spent most of my life in line organizations, and I feel I have more control over my destiny there than I do in the project. Maybe it is a personal thing for me. For example, in the project office I have had difficulty in getting the attention of people working on the project in the functional area. You're assigned people of varying capabilities from the functional areas to work for you. But whatever they're like you've got to depend on them to get the job done. If someone's not performing, you can go to his boss. But your project is just one of many to his boss. And he's in a different organization than you, so you have little direct control there. If you keep hammering on his boss's door every day, the man himself will resent it and is not going to be effective.

A recently assigned Assistant Project Manager described his point of view another way:

I was in development engineering until the middle of March of this year, so I have been on the other side of the fence doing the design and development of the engine, and always complaining about those "dirty guys over in the project office." Now I'm in the project office. When you're sitting over in development engineering, you're trying to do a good technical job—an engineer likes to get a thing perfect. The guys in development and design engineering are not looking at it from a business standpoint.

That is why you have a project office. Somebody has to make a profit for the company.

Perhaps there is more satisfaction to be found in the functional organization. The engineers find solutions to detailed technical problems themselves, while the project managers deal with problems which have to be solved by others. The project manager's satisfaction seemed to be in getting the total job done within the schedule and the budget.

Others spoke of some of the less attractive aspects of the job.

Some of the bad aspects are that it's really difficult to pinpoint responsibility and it's also hard to identify yourself with a specific accomplishment because there are so many other people involved in that accomplishment. Personal identity is sometimes hard for people in the program office to feel, so I would imagine that it is even harder for someone in a functional lab. . . .

Stress is concentrated on those people who feel responsible. I can't say where most of the stress is in the organization. It's an individual thing. I think this goes along with the difficulty in pinpointing responsibility. You either feel responsible or you don't. The very nature of the matrix distributes stress over a number of people.

The real problem we have in many of the divisions is that the work is left to the man down at the machine, and his boss is in an office somewhere and doesn't know what is really going on. You are more or less at the mercy of that particular man. If you get one that you don't have rapport with, then you have to go work on his boss to get him moving, and frequently it is difficult to get a number of things done except under extreme pressure. This is because of the buddy system. A guy down there has a friend and he will put other people's work aside to help his buddy out.

Another assistant product manager elaborated on how conflict was resolved:

Everyone in the project office thinks his is the highest priority job, and that's, of course, what he should think. But you have to look at it from a company standpoint. Recently there was a case where we had a guy over in development engineering who we needed to do some work on our program. However, they had him assigned to another spacecraft project because of a very difficult problem they were having. If he did not work on this one, it meant a large sum of money to the company. They were in some kind of a contractual situation, where if they did not start a certain test at a certain day, they would lose a lot of fees. So the program suffered a little bit.

If there is a conflict, as in this case, it gets up to the project manager. He goes over and has a chat with the laboratory manager and they have a meeting of minds about it. "What means more to the company? We know that you want this man for your project, but we need him over here." So an agreement is generally worked out.

Another assistant product manager explained some of the initial steps in setting up projects.

Generally, we go to the divisions with our requirements and our initial breakdown of the work within the program office and probably some subsystems in mind, and say, "Look, this is how I think it should be worked out—how does it look to you?" We have to compromise. I take the original cut because I know the requirements and interfaces, but then we sit down and work out the details with them.

When I go to the functional departments, I have specific people in mind that I'd like to have work for me—people who I know can do the job.

In the end we put together a team for every contract. A lot of times it's a question of their not having people available to help us. One of the department manager's concerns is who he can give me that I'll be happy with. I never ask for a hotshot to do a small project because I know they can't afford to give him to me and it wouldn't be a proper utilization of his time. If the company was shrinking instead of expanding, all this might be different. But now they aren't worried about keeping their men busy. Rather they are worried about who they can give to whom and when. Now their concern is, "When can I have this guy back to put him on something else," rather than, "I've got to keep this guy going."

Finally, some of the project managers spoke about the question of the overall organization:

It's quite an unusual atmosphere here. I don't know what your impression is, but my first impression coming from an aircraft company four years ago was that nothing was being done here. But I found as I stayed here, that it's not that way at all. People accomplish a heck of a lot more. The key to their success is their ability to recognize a problem and to put all their effort towards solving it. If any of their jobs have problems, the first thing they do is find out what they can do to correct it. Not, what to do to *hide it.*

Another senior manager for projects summarized his thinking:

I'd like to talk just a little bit about the character of the company because I think it, in a good way, influences the ability of this kind of system to work: This outfit is always *working the problem.* I have never seen anything like it. It just seems that this whole company is infused with the idea of working hard and making itself better.

It is the most self-critical place you have ever seen, and as a result, it is not stagnant. Everything is sort of continuously changing and there is always a little degree of fuzziness around. But they are all working, and not just on their own problems. A guy is just as likely to work on somebody elses's problems. It is none of his damn business, but he's doing it anyway, presumably from pure motives. He is trying to help the other guy to do better. Organizational definitions are not real rigid. Every year they get a little more so. And many of us look at that with fear and trepidation and beat them down every now and then just for fun.

But in this organization there is enough diversity and enough talent so that the organization sort of evolves as needs change. The good parts of the organization grow and prosper and the bad parts of the organization sort of don't.

You could argue that it makes for empire building but if the strong and needed parts survive, it's a valid empire.

As indicated earlier, TRW Systems had evolved over a number of years a number of educational programs that were specifically designed to facilitate the operation of the matrix. These programs grew to be considered by most as an essential part of the organization's work. The two most permanent methods employed were known as "team building" and "interface laboratories." These methods and some reactions to them are briefly described below.

Team Banding

An outside consultant described a team development session for a launch team:

TRW has a matrix organization so that any one person is a member of many systems simultaneously. He has interface with many different groups. In addition, he is continually moving from one team to another, so they need team development to get the teams off to a fast start. On a launch team, for example, you have all kinds of people that come together for a short time. There are project directors, manufacturing people, the scientists who designed the experiments, and the people who launch the bird. You have to put all of those people together into a cohesive group in a short time. At launch time they can't be worrying about an organizational chart and how their respective roles change as preparation for the launch progresses. Their relationships do change over time, but they should work that through and discuss it beforehand, not when the bird is on the pad. The concept of the organization is that you have a lot of resources and you need to regroup them in different ways as customers and contracts change. You can speed up the regrouping process by holding team-building sessions.

An SPM discussed the effect of team building on his subprogram's success:

I attribute a lot of the success our subsystem has had to our relationship with the project office and my relationship with the people who worked on it. The project office started by having off-site meetings. My project manager actually initiated it with his SPMs and some of the other APMs who were involved with us, and he chaired these meetings. We started by going to a restaurant where we had dinner and a couple of drinks and just talked to each other. But we did it under a different atmosphere. We weren't under the pressures of daily problems. The first session was more friendly than anything else—getting to know each other. In later sessions, however, we really grappled with some meaty problems. We dealt with relationships and problems directly, bringing a lot of things into the open. For example, we found out that there were a lot of problems within one of the functional areas and it helped me prevent similar things from occurring in my own area. It also helped the guys in that area clear up their own.

I was impressed enough by this experience to obtain funding for similar meetings with my own unit engineers. When we sat down, we got a lot of things squared away. It was an opportunity for everyone to talk about problems they were having by themselves and problems that others were causing. For example, when a design engineer wants to make a change, the production engineer screams at him because the change screws up his operation. Now, the production guy has a good feeling for what problems the design engineer has and vice versa, and they can talk to each other. This was all a very important part of making my people project-oriented, and I think was behind our success.

Interface Labs

As a result of the nature of the work at TRW and of the matrix, there was a great deal of interaction between the various groups in the organization. Sometimes this interaction was characterized by conflict. The personnel staff began to work on ways to help groups deal with this conflict. One such effort, the first interface lab, developed out of an experience of the Director of Product Assurance.

I came to Product Assurance from a technical organization so I knew very little of what Product Assurance was about. First, I tried to find out what our objectives were. I talked to our supervisors and I found there was a lack of morale. They thought they were second-class citizens. They were cowed by the domineering engineers and they felt inferior. I decided one of the problems was that people outside Product Assurance didn't understand us and the importance of our job. I concluded that that was easy to solve: we'd educate them. So, we set out to educate the company. We decided to call a meeting and we drew up an agenda. Then, as an afterthought, I went to see personnel to see if they had some ideas on how to train people. They just turned it around. They got me to see that rather than educating them, maybe we could find out how they really saw us and why. Well, we held an off-site meeting and we identified a lot of problems between Product Assurance and the other departments. After the meeting, we came back and started to work to correct those problems.

After this successful interface meeting, the idea caught hold and similar meetings were held by other groups. The Director of Finance, Harold Nelson, held an interface meeting between four members of his department and a number of departments that had frequent contact with Finance. The purpose of the meeting was to get feedback on how Finance was seen by others in the organization. Commenting on the effectiveness of the meeting, Nelson added:

> They were impressed that we were able to have a meeting, listen to their gripes about us and not be defensive. The impact of such meetings on individuals is tremendous. It causes people to change so these meetings are very productive.

A participant in this meeting from another department observed that, prior to the meeting, his group felt Finance was too slow in evaluating requests and that the Director and his subordinates "were too meticulous, too much like accountants." He commented that the meeting had improved the performance of Finance:

> I think Harold [Nelson] got what he was looking for, but he may have been surprised there were so many negative comments. I think there are indications that the meeting has improved things. First, Harold is easier to get a hold of now. Second, since the meeting, Harold brought in a new man to evaluate capital expenditures and he's doing a top job. He's helpful and he has speeded up the process. I think the atmosphere of the whole Finance group is changing. They are starting to think more of "we the company" and less of "us and them."

NOTE

1.   Michael Beer collaborated with us as the senior author of this chapter.

6
MATRIX PATHOLOGIES

Matrix organizations, like any other organization form, can suffer from a variety of pathologies. Because it is a relatively new form, companies that have adopted it have, of necessity, been learning on a trial and error basis. The mistakes of these pioneers can be very informative to firms that follow. In this chapter we have pulled together our observations of some of the more common problems that are associated with the use of matrix. Many of these difficulties occur in more conventional organizations but the matrix seems somewhat more vulnerable to these particular ailments. It is wise, therefore, for managers moving toward a matrix to be familiar with the diagnosis, prevention and cure of these problems. We will consider in sequence power struggles, anarchy, groupitis, collapse during economic crunch, excessive overhead, decision strangulation, sinking, layering, and navel gazing.

POWER STRUGGLES

Diagnosis. We have defined the essence of a matrix as having dual command and said that for such a form to survive there needs to be a balance of power. The idea of a power balance has always been one that seems to shift constantly. A balance of power among nations is always shifting; each party is always jockeying about, trying to gain an advantage. To the extent that the balance is operative, it will only be maintained by constant vigilance and rapid moves and countermoves. In structural terms there is an inherent instability.

The United States government, for example, is predicated on a system of balanced power between three different activities—legislative, executive, and judicial—and also by a balance between federal, state, and local rights. We speak of a system of "checks and balances" because we know that it is not enough to simply create that balance, but we must also have continual mechanisms for checking the imbalances that creep in.

The same kind of problem exists in business organizations that operate with a balance of power model: there is a constant tendency towards imbalance. So long as each group or dimension in an organization is going to maximize its own advantage vis-à-vis others, there will be a continual struggle for dominant power. Power struggle in a matrix is qualitatively different from that in a traditionally structured hierarchy because in the latter it is clearly illegitimate within the rules of the pyramid, whereas in the matrix it is a logical derivative of the ambiguity and shared power that has been built purposefully into the design. A design, therefore, that intentionally lets the boundaries of authority and responsibility overlap into different peoples' domains is bound to cause those individuals to maximize their own advantage.

Prevention.   As with nation states, corporations will find it exceedingly difficult to prevent power struggles from developing. Strength will be the best preventative. While it will not likely prevent power struggles from developing, it will prevent them from reaching destructive lengths. Strength here means the healthy power and vigor of the balancing dimensions. Friendly competition is encouraged, but all-out combat is severely punished.

Cure.   The best way to cure power struggles before they destroy the viability of matrix management and organization is to ensure that the key players on the axes of power understand that to win power absolutely is to lose performance ultimately. These key executives have to see that total victory of one dimension only ends the balance, thus finishing the duality of command, and destroying the matrix. But they must see this as an underlying principle, before and during all ensuing power struggles. The struggles will always develop. What they can only hope to prevent is that one side totally wins or loses. The key executives must therefore have a sense that they need worthy adversaries, counterparts who can match them and turn the conflict to constructive ends. For this, two things are necessary. One is that the managers always maintain an institutional point of view; that they view their struggles from a larger, shared perspective. The other is that they jointly agree to remove managers who, through weakness or whatever inability, are losing ground irrevocably; and that they replace them with the strongest available force—even if that means placing the dominant manager in the weakened part of the organization and reversing the manager's power initiatives.

Another key element in curing power struggles before they get out of hand and destroy the balance lies in the common superior to whom the power-sharing managers report. The paradox of the matrix here is that a single strong individual, one who does not share the authority that is delegated to that individual (say by the board), is needed to arbitrate among power-sharing subordinates. The top manager has many vehicles for doing this, as discussed in Chapter 4: time spent with various people, pay, promotion, direct orders, and the like. It is paradoxical because this individual, essentially, is saying to subordinates, "You *will* enjoy democracy among your ranks." What the strong individual must protect is the weak dimension in the organization, not necessarily the weak manager in charge of that dimension.

ANARCHY

Diagnosis.   Many managers who have had no firsthand familiarity with matrix organizations tend to have half-hidden fears that matrix equates with anarchy. Are these concerns based on real hazards? Certainly there exists today such a considerable number of organizations that are making successful use of matrix forms that we need not treat anarchy as a general hazard of the matrix. But are there certain conditions or major misconceptions that could lead a company into the formless confusion that we associate with the word anarchy? We cannot claim to give any conclusive answers to this question but some observations may help.

Through firsthand contact the authors know of only one example of an organization that was using a latent matrix form and that, quite literally, came apart at the seams during a rather mild economic down-turn. This organization had a fast-growth strategy that used its high stock multiple to acquire, and then completely assimilate, smaller firms in the recreation equipment field. Within a period of about six months, the firm changed from an exciting success to a dramatic disaster. Its entire manufacturing, distribution, and financial systems went out of control and left it with unfilled orders, closed factories, distress inventories, and huge debts, all at the same time. Of course, there are many possible explanations of why this happened but one perfectly reasonable explanation is that their organizational design failed under stress. What was that design?

The company essentially used a functional structure. As each small organization was brought in, its owners and general managers were encouraged to leave, and its three basic functions of sales, pro-duction, and engineering were attached to their counterparts in the parent organization. In addition a "product manager" was appointed within the overall marketing department. This individual was usually a young aggressive manager who was asked to develop a comprehen-sive marketing plan for the acquired product line that included sales forecasts, promotion plans, pricing plans, projected earnings, etc. Once this plan was okayed by top management the selected product manager was told to hustle around and make those plans come true. This is when the latent matrix came in. The product manager had to work across functional lines to try and coordinate production sched-ules, inventories, cash flow, and distribution patterns without any ex-plicit and formal agreements about these relationships. Furthermore, the product manager felt locked into the approved marketing plan. When sales slipped behind plan, the manager's response was to exhort people to try harder rather than to cut back on production runs. And once one or two things began to go wrong, there was not enough reserve in the system to keep everything else from going wrong. A power vacuum developed, followed by a winner-take-all mentality. The result was that a mild recession triggered conditions approaching anarchy.

Prevention.   We believe the lesson of this experience is loud and clear. Do not rely too much on an informal or latent matrix to coordinate critical tasks. Make such arrangements explicit so that people are in approximate agreement about who is to do what, under various circumstances. Matrix, properly understood, is anything but leaving such matters in an indefinite status. It is a definite structure and not a "free form" organization. One of the authors found evidence of this point in a recent study of five medical schools, notorious for their tendency for anarchy. The school that had moved the furthest toward developing an explicit matrix structure scored the lowest on an "anarchy index," the percentage of faculty who recognized no "boss" to whom they felt responsible for the performance of a major part of their work.

Cure.   All managers will hope that it will be unnecessary to employ a "cure" for anarchy because things will not be allowed to become that bad before adequate prevention measures are taken. However, should such a condition arise, it would call for a true crisis management response. The crisis response is really no mystery. The CEO must pull into the center all key people and critical information and must personally make all important decisions on a round-the-clock schedule until the crisis is over. Then and only then can the work be undertaken of reshaping the organization so that it can avoid any future crisis.

GROUPITIS

Diagnosis.   One way a matrix organization can get into serious trouble is by confusing matrix behavior with group decision making. The potential for such a confusion probably arises from the fact that the use of new project or business teams may be a part of evolving toward a matrix and the team idea suggests group decision making. Some decision making in groups is, of course, a perfectly sensible thing to do under many circumstances. But, if it is believed to be the essence of matrix behavior, difficulties are to be expected.

We have seen one matrix organization that had a severe case of "groupitis." Each product in this multiproduct electronics firm had a product manager and a product team comprising specialists drawn from the ranks of every functional department. So far, so good. But, the idea became prevalent in this organization that their matrix somehow required that *all* business decisions be hammered out in group meetings. To make decisions in other ways was considered illegitimate and not in the spirit of matrix operations. Now many of the needed decisions involved detailed matters with which only one or two people were regularly conversant. Yet, all team members were constrained to listen to these issues being discussed to a decision point, and even expected to participate in the discussion and influence the choice. Some individuals seemed to enjoy the steady diet of meetings and the chance to practice being a generalist. A larger number of people, however, felt their time was being wasted and preferred having the best informed people make relevant decisions without drawing them away from their specialized work. The engineers, in particular, complained that the hours they were spending in meetings were robbing them of opportunities to strengthen their special competence. They feared for their specialized identities. Beyond these personal reactions, senior managers reported disappointment about the speed and flexibility of organizational responses.

Prevention.   Because the whole idea of a matrix organization is still unfamiliar to many managers, it is understandable that it can be confused with ideas such as group decision making. The key to prevention is education. The strategic choice to move toward a matrix, as we have emphasized again and again, needs to be accompanied with a serious educational effort to clarify in the minds of all participants what it is and what it is not.

Cure.   In this case the problem came to light through interviews with professional employees in the course of reviewing the matrix approach. Once senior people had clearly diagnosed the problem, it was 90 percent cured. Top management emphatically clarified that there was nothing sacred about group decisions and that it was not sensible to have all product team members involved in all decisions. The line between individual and group matters was drawn on a straightforward basis and work proceeded in the matrix on a more economical, and responsive, basis. The concept of teamwork was put in perspective. Group decision making should be done as often as necessary, and as little as possible.

COLLAPSE DURING ECONOMIC CRUNCH

Diagnosis.   We have noticed that matrix organizations seem to blossom during periods of rapid growth and prosperity, and to be buffeted and/or cast away during periods of economic decline. Upon reflection, this is rather natural and logical. In prosperous times, firms often expand their business lines and the markets they serve. The ensuing complexity may turn them towards matrix management and organization. If these firms follow anything like a business cycle, there will be a period of two to five years before another economic crunch begins, which is more than enough time for the matrix concept to spread throughout an organization. By that time, the matrix is a familiar part of the organization, it occupies a central place in company conversations, and people have gotten rather used to it. Although there may still be some problems, the matrix seems here to stay.

When the down part of the economic cycle begins, senior management often becomes appreciably bothered by the conflict between subordinates, by the endless number of meetings and inaction, and by the slowness with which their charges respond to their poor situation. "We need decisive action" is the rallying cry. Authoritarian structures can act quickly because the spectrum of opinion need not be considered. So the top boss takes command in a different, almost but not quite forgotten, way by ramming directives down the line. There is no more time for organizational toys and tinkering. The matrix is done in.

Prevention.   Preventing this kind of collapse in the matrix is accomplished by actions that are related to general managerial excellence independent of the matrix, and that take place long before the crunch arrives. Good planning, for example, will foretell downturns in the economic cycle, for which the corporation presumably will prepare. Corporate structures should not be expected to change with standard changes in the business cycle. When management planning has been poor, however, the matrix is a readily available scapegoat. Another way of saying this is that firms which get into severe economic crunches experience the need to take drastic actions in many directions at once: cutting loose of product lines, closing offices, massive firings, drastic budget cuts, and also closer managerial controls. The matrix is often done in during this period for several reasons: it represents too great a risk; "it never really worked properly," and giving it the coup de grace can disguise the failure of implementation; the quality of decision making had not improved performance sufficiently to survive the hard times. Measures to prevent this pathology do not lie within the matrix itself so much as within improvements in basic managerial skills and in fundamental business planning. The Carlson Company case study at the end of Chapter 7 is an example of anticipation of an economic crunch and flexible response that demonstrated the strength rather than the weakness of the matrix under such conditions.

Cure.   When the matrix does collapse during an economic crunch, it is very unlikely that it can ever be resurrected. At best, the organization will go back to the pendulum days of alternating between the centralized management of the crunch period and the decentralized freedoms of more prosperous days. Top management cannot be trusted: "They said they were committed to the matrix, but at the first sign of hard times all the nice words about the advantages of the matrix turn out to be just that—nice words." If the conditioned response of a firm to hard times is anything like the above, it should not attempt the matrix in the first place. This is one pathology that requires preventive treatment. We don't know of any cure.

EXCESSIVE OVERHEAD

Diagnosis.   One of the concerns of matrix organization is that it is too costly. On the face of it, matrix organization does seem to double up on management because of their dual chains of command. This issue deserves thoughtful consideration.

The limited amount of quantified research on the subject of matrix overhead costs indicate that in the initial steps of moving to a matrix overhead costs do in fact rise, but that as a matrix matures these extra costs disappear and in fact, productivity gains appear. The experience of the authors supports this finding. In one situation we observed in some detail not only how initial overhead increases necessarily occur but also how they can inflate unnecessarily. In this instance, a large electronics firm decided to set up a new operating division in a new plant site employing the matrix design from the outset. This unique organizational experiment had a number of positive attributes but one of its problems was with overhead costs. It seems that in planning the staffing patterns for the new division, it was assumed that every functional office and every product manager's slot needed to be staffed by a full-time boss. This resulted in a relatively small division having full-time functional group managers, full-time product managers, and full-time top-level managers. With time, this top-heavy organization shook down more reasonable staffing patterns.

Prevention.   The trouble here was the assumption that each box required a full-time incumbent. This is much less apt to happen when an organization evolves gradually from a conventional design into a matrix. In these circumstances managers frequently double up in roles. Managers can be asked to perform both as functional managers and as product managers. While this technique can be justified on a temporary basis and as a transition strategy, it also has its hazards as we saw in the discussion of power struggles above. A safer route is to assign managers to two roles on the same side of the matrix, dual functional roles, or dual product management roles.

In the last analysis, no organization would adopt a matrix without the longer run expectation that, at a given level of quality output, the costs of production would be less than with other organizational forms. In what way can such economies be achieved? We would say that the potential economies come from two general sources: fewer bad decisions and less featherbedding. The process by which the matrix can potentially improve quality of business decisions is taken up in a number of other places in this book. Essentially it happens because the matrix helps bring the needed information and emphasis to bear on critical decisions in a timely fashion. This is the more important potential for economies. The second source, no featherbedding, is less obvious but potentially of significance. How can it work?

Cure.   Perhaps the clearest example of matrix potential in reducing redundancies in human resources can be drawn from the way the matrix has been used in some consulting organizations. The consulting firms we have in mind (see Chapter 7) usually set up a matrix of functional specialist managers against client or account managers. The rank and file consultants are grouped with their fellow specialists but they are available for assignment to projects under the leadership of account or engagement managers. From an accounting standpoint, consultants are, therefore, only working for account or engagement managers when their time is directly billable to clients—otherwise, their time is charged against the budget of their functional manager.

This nonbillable time is, therefore, very conspicuous, both to department and to individual consultants. Of course, not all time charged to functional departments should be thought of as redundant. Necessary background studies and research work would normally be carried out in this category as well as the inevitable turnaround time between assignments. But such work can be budgeted in advance and variances from budget subjected to careful scrutiny. As one senior manager put it, "There is no place to hide in a matrix organization." This fact makes for a demanding organization. Rank and file people do feel a clear-cut pressure to produce. In fact, care must be taken by senior people to keep such pressures from becoming too strong for the long-term good of both the people involved and the organization. It is perfectly possible to get too much as well as too little of such pressure. A creative tension is sought.

DECISION STRANGULATION

Can moving into a matrix lead to the strangulation of the decision process? into endless delays for debate, for clearing with everybody in sight? Will decisions, no matter how well thought through, be made too late to be of use? Will too many have veto power? Will all bold initiatives be watered down by too many cooks? Such conditions can arise in a matrix. We have in mind three typical situations, each calling for slightly different preventive action and cure.

(1) Constant Clearing.   In one firm we know of, the various functional specialists who had been selected to report to a second boss, the product manager, picked up the idea that they had to clear all issues with their functional bosses before agreeing to product decisions. This meant that every issue had to be discussed in at least two meetings. The first meeting could only review the facts of the issue, which was then tabled until all the specialists cleared the matter with their functional bosses—who by this process were each given a de facto veto over product decisions. This impossible clearing procedure represented, in our view, a failure of delegation, not of matrix. One needs to ask why the functional specialists could not be trusted to act on the spot in regard to most product decisions in ways that would be consistent with the general guidelines of their functional departments. Either the specialists were poorly selected, too inexperienced, and badly in-

formed, or their superiors were lacking in a workable degree of trust of one another. In either case this problem, and its prevention and cure, needs to be addressed directly without making a scapegoat of the matrix idea.

(2) Escalation of conflict.   Another possible source of decision strangulation in matrix organizations comes from escalation of conflict. This source of possible problems, derivative from frequent or constant referral up the dual chain of command, is one of those most frequently raised by managers unfamiliar with the matrix. They are highly aware that one advantage of the conventional single chain of command is the simple fact that two disagreeing peers can go to their shared boss for a resolution. They look at a matrix and realize that their nearest shared boss is the CEO, who could be five or six echelons up. They realize that not too many problems can be pushed up there for resolution without creating the ultimate in information overload. So, will not the inevitable disagreements lead to a tremendous pile up of unresolved conflict? Certainly this can happen in a malfunctioning matrix. Whether or not it does happen depends primarily on the depth of understanding that exists about required matrix behavior on the part of managers up the line in the dual structure. Let us envision the following scene: a 2-boss manager gets sharply conflicting instructions from the product boss and the functional boss. The manager tries to reconcile conflicting instructions without success, and quite probably asks for a session with the two bosses to resolve the matter. The three people meet and the discussion bogs down, neither boss gives way and no resolution is reached. The two bosses then appeal the problem up a level to their respective superiors in each of the two chains of command. This is the critical step. If the two superiors properly understood matrix behavior, they would first ascertain whether the dispute reflects an unresolved broader policy issue. If it doesn't, they will see that it is their proper role to teach their subordinates to resolve the problem themselves—not to solve it for them. In short, they would not let the unresolved problem escalate but rather would force it back to the proper level for solution and insist that it be done promptly. Often, conflict cannot be resolved; it can, however, be managed. Such behavior is clearly essential if the matrix is to work. Any other course of action would represent a failure to comprehend the essential nature of the design.

(3) Unilateral style.   A third possible reason for a growing sense of decision strangulation in a matrix system can arise from a very different source—personal style preferences. Some managers have the feeling they are not truly managing if they are not in a position to make crisp, unilateral decisions. They identify decisive actions with leadership. They develop strong feelings of frustration when they have to engage in carefully reasoned debates about the wisdom of what they want to do. This feeling can develop even in regard to a business problem whose resolution will vitally effect functions other than their own. Such interdependencies are characteristic of many product-line decisions, especially in firms that are experiencing critical dual pressure from the marketplace and from advancing technology. The matrix offers a way to deliberately induce simultaneous decision making between two or more perspectives. It fosters bilateral rather than unilateral decisions. If managers start feeling threatened by bilateral decision making, they are certain to be unhappy in a matrix organization. In such cases the strangulation is in the eye of the beholder. Such people must rework their concept of decision making or look for employment in a nonmatrix organization.

SINKING

Diagnosis.   There seems to be some difficulty in keeping the matrix viable at the corporate or institutional level, and a corresponding tendency for it to sink down to lower levels in the organization (group, division) where it survives and thrives. For example, a company sets up a matrix between its basic functions and product groups. The product managers never truly relinquish their complete control, and the matrix fails to take hold at the corporate level. But one or two of the managers find the idea useful within their divisions; their own functional specialists and project leaders can share the power they have delegated, and the design is effective within subunits of the corporation. For example, Dow Chemical's attempt to maintain the product/geography balance failed at the corporate level but held within the geographic areas for several years.

Sinking may occur for two reasons. Senior management has either not understood or been unable to implement the concept as well as those beneath them. Or the matrix has found its appropriate place.

When the former is the case, it is likely to occur in conjunction with other pathologies, particularly power struggles. Senior executives may have played around with the idea, and only one or a few got convinced. There is a danger in such cases because some of those at the top espoused a philosophy and method they could not employ themselves while others were able to show that it did work (so long as it was their subordinates, not they, who had to share authority).

Prevention.   Sinking can be prevented by adequate conceptualization. What this means is that a corporation must think through which dimensions it must balance, and at what level of aggregation. For example, must all business units matrix with central functional departments? If the answer is no, then some will operate as product divisions with the traditional pyramid of command, while others will share functional services in a partial matrix. Sinking should be prevented only when it is indicative of disintegration or decomposition of an appropriate design.

Cure.   As often as not, sinking is a healthy phenomenon. It may simply be reflecting periods of adjustment. Before matrix management can be put in place (run smoothly), it must be put in the right place (proper location). This can be better thought of as settling, than as sinking. Settling into parts of an organization, rather than providing the framework for the total institution, is likely to occur during the early evolution of a matrix.

Settling also leads to units of manageable matrix size. One of us spoke at a conference with William Goggin who described his success with the matrix at Dow Corning. The main question put to us by the several large corporations attending was: "That sounds great for a $250 million company with a few thousand employees, but can it work for a $2-3 billion company with 50,000 employees? Its entire company is the size of one of our divisions." Matrix management and organization do seem to function better when no more than 500 managers are involved in matrix relationships. Within that size the people who need to coordinate regularly are able to do so through communication networks that are based on personal relations. Remember, that even in a firm of 50,000 only 500-1500 managers will be in the matrix; so in a firm with 500 employees, only about 50 will be involved in dual reporting lines.

Whatever the size of the unit in which the matrix operates, the important action is to have made the correct conceptualization. Then if settling occurs, it should be seen as a cure itself and not as a pathology. Sinking may be a healthy self-adjustment that suggests the organization's capacity to correct its own errors.

A Corollary.   Sinking is the tendency of a matrix to settle down to lower levels in an organization. We also want to call attention to an opposite or corollary tendency: *floating.* This is the rise of a dual management to successively higher levels in the organization. Wherever we have seen floating occur, it has been experienced as a healthy response to some very real environmental pressures. Whatever problems are related to it are only those that have retarded the float, prevented it from occurring sooner. And the retarding forces we have seen are power struggles in the organization. Functions, products, services, and markets that rise to higher levels of integration in the design appear to be getting proper recognition. Because of the balance in the matrix, this float is fundamentally different from the centralization of the same element in the traditional pyramid structure.

LAYERING

Diagnosis.   Layering is a phenomenon in which matrices within matrices are found, often cascading down through several levels and across several divisions in an organization. This phenomenon may or may not be pathological. We want to emphasize that this may be a rational and logical development of the matrix, but we include it briefly here because it sometimes creates more problems than it solves. In terms of the metaphor we have used in this chapter, layering is a pathology if the matrix begins to metastasize. When this occurs, organization charts begin to look like blueprints for a complex electronic machine, relationships become unnecessarily complex, and the matrix form may become more of a burden than it is worth.

Prevention and Cure.   The best remedies are adequate conceptualization and the reduction of power struggles. We have seen a few cases where one dimension of the matrix was clearly losing power to the other, so it adapted an "if you can't beat 'em, join 'em" philosophy and grew a matrix within the dimension. The best defense was a good offense, or so it seemed. A product overlay, for example, developed its own functional expertise distinct from the functional units at the next level up. In two other cases international divisions each created its own matrix by adding business managers as an overlay to its geographic format, without reconciling the difference between these and the ones who run the domestic product/service groups. Adequate conceptualization in each case would probably have simplified the organization design and eliminated the layering. In each case the layering occurred because of unacknowledged power maneuvers. This unhealthy state can best be cured when the matrix managers are relatively well balanced, when a steady state exists, and when no one is either imminently threatened or pushing hard toward a power goal.

Matrix design is complex enough without adding to it because of the need to balance power. Remember, a well-conceptualized matrix is bound to be less complex and easier to manage than one that is not logically organized. Layering is frequently the result of the logic of power rather than the logic of design.

NAVEL GAZING

Diagnosis.   Because the matrix is found in organizations in which there is considerable interdependence of people and tasks, and because the matrix demands considerable negotiating skills on the part of its members, there is sometimes a tendency to get absorbed in internal relations at the expense of paying attention to the world outside the organization, particularly to clients. When this happens, more energies are spent by the organization on ironing out their own disputes than on serving their customers. The outward focus disappears because the short-term demands of daily working life have yet to be worked through. The navel gazing is not at all a matter of lethargy, rather it can become a heated fraternal love/hate affair among the members. This occurs more frequently in the early phases of a matrix, when the new behaviors are being learned, than in matrices that have been operative for a few years.

Prevention.   Whatever other pathologies develop in a matrix, attention to their cure is bound to increase the internal focus of the members; so prevention of other pathologies will certainly reduce the likelihood of this one occurring. Awareness of the tendency will also help.

Since one dimension of the organization (the output dimension) generally has a more external focus than the other (the resource dimension), the responsibility for preventing an excessive introspection is not equally distributed. A strong marketing orientation, probably predating the matrix, is the best preventive of all.

Cure. If the players in the matrix are navel gazing, paying excessive attention to internal affairs at the expense of adequate service to clients, the first step in the cure is to become aware of the effects. Are customers complaining a lot, or at least more than usual? Agreement should be reached that internal conflict must be confronted, but that it is secondary to maintaining effective relationships externally. The ultimate cure, however, lies in working through the internal conflicts without becoming obsessed by them. Navel gazing generally occurs when the matrix has been fully initiated, but not yet debugged. People accept it, but they are engrossed in figuring out how to make it work. The problem partially then can be confronted head-on, and partially by treating the inward focus as a symptom of the underlying issue: how to institutionalize matrix relationships so that they become familiar and comfortable routines. Finally, it must always be remembered that any form of organization is only a means to an end, and should never become an end in itself.

CASE STUDY: DIAMOND INSTRUMENT COMPANY

By the summer of 1974 Diamond Instrument had accumulated over four years of experience in managing a matrix organization. The company had adopted the matrix as a way to achieve the objectives of its different business units while simultaneously maintaining its specialized competence, particularly in engineering. However, Diamond Instrument had also discovered that the matrix raised a whole set of difficult problems around the design and implementation of the organization and, in particular, around the allocation of people between business units and functional groups.

The Company

Diamond Instrument (DI) Company was a privately owned corporation engaged in the design and manufacture of electronic instrumentation for measurement and for analysis. In 1974 the company was approaching the 60th anniversary of its founding. Annual sales were just under $45 million. Corporate offices were located in Evanston, Illinois, with manufacturing plants in Massachusetts, California, Florida, and Paris, France. Marketing offices were located in nine states and in five European countries and Canada. Service operations were handled by 87 different locations throughout the world.

Diamond Instrument had long enjoyed a reputation as an "engineer's company." The company developed excellent engineers with talents in numerous areas and these engineers kept producing new product ideas related to their special expertise. A very diverse product line evolved in the company. Sales personnel, however, were seldom included in product planning.

All of DI's senior management had risen through engineering. They knew only too well the importance of DI's technical competence, a competence which had recently expanded into several new technological domains. They now boasted expertise in fields such as computer software systems, microcircuit design and fabrication, and automated systems design. They were resolved to maintain these newly acquired but different skills.

However, DI's management was also aware that extensive commonalities existed across these different specialties. Significant economies of scale could be realized in the common production and shared marketing of similar products. It was this realization coupled with the knowledge that something had to be done to contain the proliferating numbers of new products and to relate them more effectively to real market needs that drove them to adopt a matrix design.

Implementing the Matrix at the Top

During most of its corporate history, DI had employed a straightforward functional form of organization. DI began the process of evolving beyond this first phase and toward a matrix in October, 1970, when the then president, Dr. H. W. Townsend, distinguished between functional organization and product–line organization, referring to the latter as business–unit organization. (See Fig. 6.1.)

In this phase, multiple roles were filled by many of the top managers. For example, Robert Northrup was both senior vice-president for the business and subsidiaries plus functional manager of the marketing area. W.B. Simpson was vice-president in charge of business units and simultaneously functional head of engineering. Similar dual roles, either between product and functions or between corporate office and function, existed for three other senior managers. In addition, the direction of each of the business units was handled by a manage-

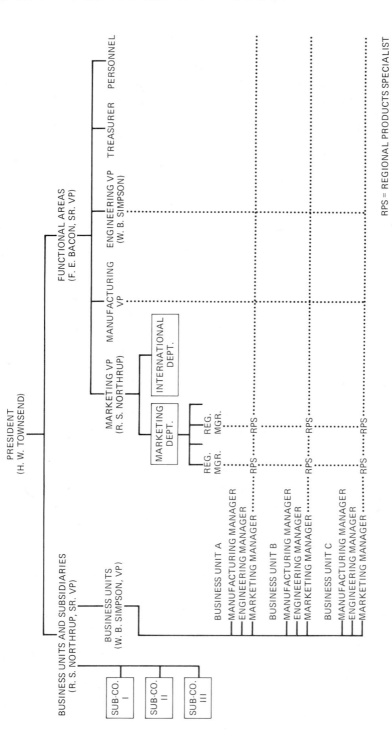

Fig. 6.1 Diamond Instrument Company's partial organization chart (case-writer's conception), October 1970.

RPS = REGIONAL PRODUCTS SPECIALIST

ment team composed of one representative from each of the three basic functions of marketing, manufacturing, and engineering and was supported by product planning personnel and by product specialists. Responsibility and authority within the team were not clearly spelled out. The needs of cooperation and coordination around specific product-market goals, however, appeared to DI management to be more important than such issues.

This early form of Diamond Instrument's matrix organization was in operation through the 1970–1971 electronics industry recession. By 1972, DI found itself coping with not only an unfavorable economic environment but some basic frictions in the implementation of the matrix design. The new business groups were beginning to flourish but the top level of management was experiencing difficulties in changing behavior to the new requirements.

The problem centered around the division of labor between the two senior vice-presidents, one in charge of functional areas and one in charge of business units. Most of the problems that arose had both "functional" and "business" aspects to them. Optimum trade-off decisions had to involve both senior vice-presidents, as organizational equals, and get their detailed joint agreement. Unfortunately, the two senior VPs differed substantially in their approaches to problems, in their styles of problem solving and delegation, and in their understandings of their own and each other's responsibilities and authority. These conditions had been left to be worked out over time as experience with the new matrix organization accumulated. The president, not wishing to take sides and desiring to encourage the personal development of his two key subordinates, was reluctant to step in and make decisions himself. The end result was that most important decisions were greatly delayed; and the company became caught up in various types of intramatrix power struggles.

The president finally decided that "one individual should be given responsibility and authority to manage all the company operations, on both sides of the matrix." On December 29, 1972, R.S. Northrup became president of the company with responsibility for all the company's operations. The senior vice-president who had been in charge of the functional areas retired from the company after 32 years service. Dr. Townsend became chairman of the board after 37 years of service with Diamond Instrument. This constituted the third phase in the life cycle of the matrix.

The fourth phase occurred some six months later when a single individual was appointed from each of the market/product teams to become general manager of each of the business units. The new general managers reported directly to the president. (See Fig. 6.2.) None of the general managers held a simultaneous functional post.

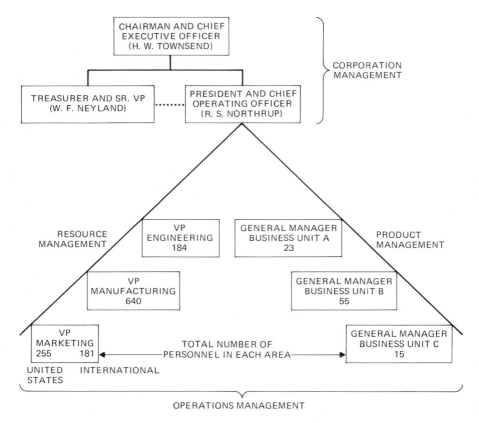

Fig. 6.2   Diamond Instrument Company's top-management organization, July 1973.

Lower Level Implementation

One of the first "2-boss jobs" DI created was that of regional product specialist (RPS). The RPS's job was the one that the company really cut its conceptual teeth on in coming to understand and clarify the multiple reporting roles in the matrix organization. The RPS was to be jointly managed by a regional manager and a business marketing manager.

An internal memo to the marketing department by the top managers of the relevant chains of command explained how this new RPS role was to function. They expected that the new organization form would create "problems of confusion, conflict, and turmoil." Nevertheless, they chose to use the implementation process as a vehicle for understanding the interplay between functionally oriented managers and business-oriented managers, and to improve their overall understanding of how the matrix worked. Their memo extensively described their thinking on issues of RPS selection, training, phase-in, performance-rating and appraisal, salaries and bonuses, and communications and coordination. Some of this thinking is reproduced in Fig. 6.3.

The memo also covered specific expectations of the RPS and his two superiors with respect to forecasts, quotas, commissions, budgets, and expenses. For example:

> There will be separate forecasts and quotas by product line and by territory, related to individual performance. The business marketing managers and the regional managers will discuss their forecasts and quotas, a practice which will substantially improve both forecasts, through "cross" education concerning new product and marketing plans and regional and customer conditions and expectations. However, it is generally agreed that the final totals need not coincide, the most important factor being that *goals exist* along *both* "axes" of the matrix and be *totally accepted by those involved as their own.*

Reassessing the Matrix

Despite the attempts to learn from the RPS 2-boss jobs, the implementation of the matrix within manufacturing, engineering, and marketing continued to follow a bumpy path. Initially, all human resources were physically and structurally located in the functional units. In time the intense interdependencies with the business units resulted in many functional personnel shifting physically to business unit locations. In some cases, personnel were also administratively transferred into the business units when the particular business manager presented a strong argument for the need for total control over the particular individual's work or for exclusive access to the individual's particular competence. The people allocation problem became increasingly confused as the balance shifted and the business units became stronger.

In September 1973, DI's top-management group held an off-site, three-day meeting to discuss DI's problems and opportunities. Among other issues, they took specific decisions to clarify organizational relationships and to set up procedures to deal with conflict: "An effective organization is not one *without* conflict, it is one with the ability to *deal* with conflict." Two decisions, in particular, involved the allocation of people:

- Key people needed for product definition will report administratively to business general managers. Specifically, mechanical engineering capability will be added to both Business Unit A and B.

Supervision of Performance of RPS and Performance Appraisal

The regional managers and the business marketing managers will be interested in somewhat different, though overlapping, aspects of the RPSs' performance. Each will supervise and appraise along his own lines of interest, but with frequent discussions to avert conflicting guidance. Everyone recognizes that there is a natural potential for conflict or playing one against another, and that such possibilities must be consciously averted by all involved.

Salary and Bonus Performance Ratings of RPSs

The business marketing managers and the regional managers will have different factors for judging and rating RPS performance. (Teamwork performance relating to *all* of DI's activities will be a major factor.) Each manager will make his ratings, then they will discuss them (valuable communication will result), after which each separately will decide his own ratings for each RPS, subject to the usual averaging constraints within his grade-groups. The two separate ratings for each RPS will then be averaged for "plugging into" the salary-structure system. *Both* managers will receive the salary data for the RPSs they supervise.

Raises for RPSs

The regional managers and the business marketing managers will discuss and agree on the timing and amount of RPS raise recommendations and will coordinate in telling the RPSs about their approved raises. Each RPS will always be told that *both* his bosses are rewarding his performance.

Fig. 6.3   Diamond Instrument company—regional product specialists.

Key people needed for specialization and development will report administratively to the vice-president for engineering. No mass changes in reporting will be made to implement this policy; rather, as new decisions for assignment of personal come up, the objective stated above will be used as a basis of assignment.

People Allocation Problems Lower Down

The decision to allocate "key people" to business units responded to some of the concerns of the business general managers but, in turn,

Communication

Communication in management has always been important. The matrix system makes it *more* important, but, at the same time, explicitly *calls for it* in day-to-day procedures, as is clear from the foregoing. The additional communication will pay off in the form of better decisions, especially in important matters where much judgment is required, such as personnel ratings and forecasts.

Cooperation

Cooperation in management has also always been important and appears in some form as an explicit factor in most personnel rating systems. The matrix system makes it *vital*, as is clear also from the foregoing. There can be fewer independent, unilateral decisions, and strongly influenced individual decisions must become the rule in the interlocking areas. We must all consciously work for a healthy blend of our naturally "orthogonal" views in order to avoid working at cross purposes. This style of cooperation requires that we all increase our sensitivity to, respect for, and response to the problems and views of others. These qualities have always been of first-rank importance at the highest levels of any corporate management; successful operation of a management matrix elevates them to first-rank importance much deeper into the hierarchies.

Just as DI must adapt its organization to a changing external environment, the key managers must, in turn, adapt their methods of operation to a changing internal environment.

Fig. 6.3 (continued)

created difficulties for the functional managers. One persistent problem was the relationship of the systems software people in functional engineering to the business product development engineers. The difficulties were in some ways unique to hardware/software interface problems and to the fact that software was a flexible form of technology. They were, however, also a reflection of the conflict between functional departments and business units.

Stan Serkes was the group leader for Software Systems in the engineering department with a staff of 15 graduate engineers plus a secretary and a technician. Stan Serkes talked about the hardware/software split.

DI has been a hardware company over the past 50 years. Computers have come in only in the last 10 years. We're still fighting for recognition; we're having to earn our place. Consequently, we feel that programming has been subjugated. It's something that's called in at the end. It's not something that's been consulted early enough in the concept.

The problems lie in two areas. One is seeing decisions made in the early stages of conceptualizing and planning a new product that we could have told them were bad decisions and that we could have showed them how to make more properly. These decisions were made and we weren't consulted. Then, there's the problem of working with people in a different area. The hardware engineers belong to Business Unit A engineering. The software people belong to my software group in functional engineering. Bringing them together into a harmonious team is awfully difficult. Furthermore, some of us are separated physically. We don't work jointly. They go off and optimize the hardware and we go off and optimize the software, and then we try to jam both together to make them work.

One of the biggest problems as far as we are concerned is that the main part of the engineering department, the circuit design engineers— the people doing creative work—have been divided up and assigned to business units. I consider that the people who design software are doing just as creative work as the design engineers and should be on an equal footing with them. One of the ways to do this, since the split up of engineering seems to be irrevocable, is to split us up too. The other way to go is to see a reassembly of the engineering organization. I'm saying that my group should be divided up and parceled out to the business units as needed, or all engineering should be drawn back together again so that we all belong to the same organization and can work together. Recently, two of my people were transferred to the business units to help them understand the software problems.

Don Richards, the engineering manager for Business Unit A, used 14 of the 15 systems software engineers. They were divided up between the four major projects within the unit. However, the degree of interdependence between the business project management and the software engineers varied considerably. Richards had some comments on the hardware/software interface:

If I had the capability in software that I have in hardware, I'd be able to make the necessary compromises more effectively in my own mind. Because we don't have the capability in our own business engineering group, we had to rely on other people to make that trade-off. I don't feel that the trade-offs were made correctly. I feel that they were incorrect because the matrix organization gets people in the functional area to be all concerned with technical and professional excellence. That's their goal, rather than satisfaction of the marketplace. They spend much of their effort honing their tools, learning about programming, and not learning about our business. As a result, when it comes to making business compromises, we end up on different sides of the street more often than not.

I've done a great deal of thinking—is it the system or is it the personalities? I think my time frame may be a contributing factor. I think in terms of here and now, of let's get things done. They think in terms of—well, let's think it out first, then we'll decide how to do it. This is a very fast-moving field. We've got to have something now. We've got to be able to ship it. And then come back and make it better the next time around. It's in this difference that many of our tensions built up.

Some of them work well with our changing objectives. For others it hasn't worked that well. I think it's because the people involved tend to be very, very thorough in analysis of the problem. When they've thoroughly analyzed it, they come back and then have a new set of objectives thrown at them. They have to start all over again. After two or three rounds of this they tend to get very upset and they withdraw. They don't participate. They wait for us to stop changing our mind. As a result communications break down.

Software is so important to our product. It's essential to it. I think it's foolish for the business units to be without it. I don't think it's a pooled thing. I've taken a very aggressive stance. When I get authorization to hire, I've been hiring computer scientists who at least have a good understanding of software.

Serkes felt that the matrix structure was a large part of the problem. He felt that DI had started to divisionalize but hadn't gone all the way and was calling it a matrix. That was central to the problem. He cited a specific matrix conflict.

We did an exercise recently about a conflict between one of my programmers and a hardware engineer over how an interface for a piece of equipment should be designed. The programmer says, "If you put in one more integrated circuit package, my code becomes 50 instructions shorter." And the other guy says, "No. My boards are already designed and I'm not going to make any more changes. You fix your software." We ran it up through four levels and still found no one in common and by that time we'd reached the level where the vice-president for engineering and the business manager were both reporting to Bob Northrup. Now here's a problem with the matrix. My boss says "You could settle that. You don't have to go to Northrup with every one of your little squabbles. Talk it over." I say this makes it inefficient. We have to persuade and cajole and harangue and hammer out compromises with meeting after meeting and people blowing smoke in each other's eyes. Why can't we just have some kind of a tree-shaped organization where somebody makes a decision and says the buck stops here?

Richards was less inclined to blame the matrix.

Matrix management isn't the cause of the problem. Organization according to matrix could work. Matrix is another layer of management rather than the form of management. There's still a lot of the old management layer left. As a result, there's struggle and a lack of feedback in the matrix management which makes it cumbersome.

When there's a problem, one would expect the matrix management system to create tension at that intersection and drive people towards a solution to that problem. In the case of the programmers and our development efforts here, the problem was a lack of communications with respect to control over tasks. Ongoingness of the problem says that the tension never got to the point where people felt that it needed to be really resolved. It was always a short-term solution.

Divisionalization

In March 1975, President Northrup created a new Environmental and Medical division out of Business Unit B plus one of the subsidiary organizations. The new division was relocated into a separate physical facility of its own. The division viewed many of its product marketing activities as unique and felt that it was justified in developing its own sales force to better serve its own products. It also obtained agreement from the president to perform its own manufacturing on all new products.

In May 1975, the Electronic Instrument Division was formed and headed by the person who was until then the functional vice-president of engineering. Most of the Business Unit C activities fell under this new division. At the same time, Business Unit A was reorganized into the Test Systems Division and its marketing organization extensively restructured to rationalize some overlaps with the functional marketing organization. The software or programming function in engineering was transferred to the Test Systems Division while most of the remaining engineering functions were absorbed into the manufacturing organization. Engineering, as a central function, was eliminated.

By the end of June 1975, Diamond Instrument had changed from a functionally dominated matrix organization to a divisionalized structure with two large centralized functions (Marketing and Manufacturing) that served each of the different divisions in different degrees. The matrix design remained at an overarching level, but the trend toward complete divisionalization appeared relentless and the outcome inevitable.

NOTE

1.   C.J. Middleton, 1967. How to set up a project organization. *Harvard Business Review,* March/April.

7
MATRIX BEYOND INDUSTRY

INTRODUCTION

Although matrix organization has been most fully developed in industry, it has also been utilized in a variety of service organizations, in both the private and the public sectors. In each instance, pressures from two or more environments force the organizations to consider the balance of power model. These environmental dimensions are always (1) geography, (2) specialized functions or services, and/or (3) distinguishable clients. Some of these businesses operate in very uncertain environments, and others find that expansion has created a web among its interdependent parts. Still, few of them turn to the matrix until their size compels them to rationalize and coordinate their far-flung mix of activities better.

The purpose of this chapter is to sketch for the reader the great variety of fields and organizations beyond industry in which the matrix may be found or considered for use. We will examine organizations in both the profit and not-for-profit sectors including: insurance, consulting firms, CPA and law firms, securities, banks, retailing, construction and real estate, hospitals, universities, and government agencies at local, federal, and international levels.

INSURANCE

The insurance business has hardly been known as a bellwether for innovative organization, but the example of one innovative insurance company that used the matrix for years will illustrate its utility and potential in this field. The firm is Skandia. Based in Stockholm, it is one of the largest European insurance firms, with 1975 sales of $912 million. The matrix was adopted in 1968, at a time when the firm was led by Pehr Gustaf Gyllenhammer. Gyllenhammer is now president of Volvo, and is well known for his innovations in the organization of work, which began earlier in Skandia.

Skandia turned to the matrix as an organizational response to a number of complex environmental changes. Increased social security benefits created compulsory life, health, and accident insurance requirements in Sweden which forced Skandia to diversify its offerings to the public. There was also considerable foreign, as well as domestic, competition and the electronic data-processing revolution created pressures to secure economies of scale. These forces triggered a series of mergers among five firms, which in turn brought about the need to establish one set of corporatewide policies that could be coordinated throughout the Skandia system.

The matrix was created in order to give balance both to the diversified product (insurance) lines and to the zonal offices that served the public. The latter were profit centers that had authority for all daily operations in all businesses in their area. Their activities included selling, service, local promotion, claims adjustment, and the administration of policies held by clients. They provided close and quick customer service. The zone managers controlled their own resources and, in the largest zone, this meant 1200 people, half in selling and half in claims settlement and administrative support. Within the selling function, it was the zone manager's job to decide how many salesmen would specialize by product and how many would cover the general range. The salesforce, left to its own devices preferred to sell only in the high commission policies, such as life insurance.

While the zonal managers were located in their regions, the product managers were centralized in the headquarters office. Their job was a combination of product design, development, and manufacture. They determined the proper product/market mix and established guidelines for risk assessment and selection. They also engaged in underwriting, market analysis and forecasting, and loss prevention programs. These were small groups in contrast to the zonal offices, and the largest numbered only 44 people.

The product managers competed for the time and resources of all zonal staff to sell their products, and they had to stay close to the marketing activities or they found themselves promoting too frequent change of policies for the zones to handle well.

The product groups included industrial, homeowners, motor, marine, life, and group insurance. Originally there were three zones, but the one that included Stockholm was so powerful that it was split in half to create the necessary power balance both among the zones themselves and between the zones and the product dimension. A corporate planning office played a coordinating and balancing role between the two. (See Fig. 7.1.)

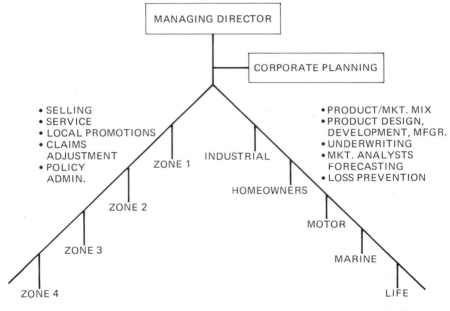

Fig. 7.1   Skandia organization, 1976, with product and zone responsibilities.

The matrix took several years to evolve and mature. During that time a definite culture grew up in which the matrix was seen as the organizational embodiment of the industrial democracy movement that was and is quite powerful in Sweden. This culture emphasized active and open exchange of views, a minimum of hierarchy, and a maximum of freedom within the workplace. The matrix was merely the organizational form for the new ethos. Architecturally, the old corridors and cubicles were replaced by "office landscaping" around "family groupings." Similarly, the company newsletter became a vehicle for lively debates on national political issues, and not the common pap of most house organs. Of course, personnel at lower levels, who constituted the vast majority in numbers, had little or no sense of the matrix—that it existed or how it functioned—but middle and senior management debated it continuously.

In the early days of the Skandia matrix, the product managers were little more than central staff groups. As their influence grew, however, the power balance became more of a reality. Around 1972–1973 the matrix seemed mature and stable. In 1975, after six years in the matrix, the product groups became market units. Although it was not seen at the time, this signaled an important shift in the power balance away from the zones. The traditional grouping of products—in accordance with objects of events to be insured—shifted to a design based on homogenous consumer groups. A personal non-life group was created, for example, to cover personal risks within the home, country house, travel, car, and boat.

During the middle 1970s the information-processing systems were maturing along with the structure. A customer could now go to a local office and get instant claim settlement through a terminal hooked up with the central data bank, instead of relying on slower personal client–zone relations of the past.

These shifts in technology and power prompted the creation of a task force, in mid-1976, to reassess the matrix. Composed of members of both sides of the matrix, together with corporate planning personnel, the outcome was to formally abandon the matrix in favor of marketing divisions, with the zones providing an administrative infrastructure. As one of their managers said at the disbanding, "The matrix is dead, long live the matrix."

The Skandia example, thus, is doubly useful. It demonstrates how insurance firms can benefit from the model, and it also emphasizes that in some cases it may be more appropriate and/or useful to view the matrix as a transitional form of organization rather than as an ultimate design. In Skandia it had utility for integrating diverse companies after a merger, and for supporting a more open managerial climate. During its eight years of existence it was clearly more than a verbal redrawing of old staff-line relations. It remains to be seen whether the passage away from the matrix will come to be seen as a pathology towards more closed and hierarchical management, or as a natural evolution to a simpler structure for a continued humane and productive environment.

CONSULTING

Consulting firms are very varied in the kind of activities they engage in, and the matrix has definite applicability to some, while none at all to others. The kinds of consulting firms that utilize aspects of matrix organization and management are those that engage in project work with clients that involves several specialized services offered by the firm that are interrelated from the client's perspective.

A consulting firm that does psychological testing, organization design, market research, or develops control systems might not turn to this format. A firm that contracts with a developing nation for a community development project, however, will likely assemble a team of its consultants drawn from various groups in the firm. These may include specialists in engineering, economics, health systems, agronomy, and management/administration. These kinds of projects might last two or three years during which time the consulting team works in a highly interdependent and semipermanent manner. There is likely to be a senior project leader coordinating the various activities. At the end of the project, however, each specialist will return to his or her group within the firm for reassignment, and during the life of the project will maintain contact with the specialist group. The latter is a permanent part of the consulting firm, even if it is formally set up as a subsidiary, and it will maintain excellence in its specific activity, see to it that its members are trained and developed (partially through rotational assignments to projects), and will also generate business for itself and possibly other related departments.

One consulting firm had been profitable but conservative. Their clients knew they would turn down assignments unless they had the expertise to do the job well. They hired good people, groomed them slowly, and gave them complete freedom in individual client-consulting relationships. They never sought related business from current clients for other parts of their firm. Organizationally, they were highly decentralized into five consulting departments: policy, marketing, administration, finance, production/engineering. By the mid-1970s they had a solid profit picture, but slow growth and a declining market share. A new president was brought in. To prevent a further slide, he was faced with a strategic choice between expanding the line of services offered to current clients or seeking new clients while remaining specialized in their client contacts. Increasing the number of new clients would call for aggressive marketing and would go against the traditional culture that did not actively seek new business, but it would be minimally disruptive to the internal organization of the firm. The full-line strategy that offered a variety of services to current and established clients would not tamper with the public image of the firm, but it would complicate the internal organization. This latter approach would probably lead to account management, internal integrators, and a higher degree of coordination among the five groups. The president is still pondering the trade-offs, but if he chooses the latter, he knows he will find himself shifting into a matrix framework.

CPA FIRMS

In the early 1960s the CPA field was experiencing a number of changes. The number of firms was consolidating, and small local firms became branch offices of larger national companies. There was a tightening up on the quality of reporting demanded by government agencies (SEC, IRS), and hence by "industry" associations such as the American Institute of CPAs and the Accounting Principles Board. An increase in third-party liability litigation against accountants necessitated better quality control for services offered. Larger CPA firms, particularly the Big 8, responded to these changes not only by improving quality but also by expanding in both the types of services provided and in their geographic coverage.

The internal organization of CPA firms was affected by this increased complexity. A simple and common form of organization was based on one-to-one relations between individual accountants and their clients. Another way of organizing, common to large CPA firms, was around the specialized services offered to clients: auditing and accounting, tax, management control, diagnostics and consulting. (These services are not to be confused with internal functions such as personnel and administration of the CPA firm itself.) The type of client serviced, offered another manner of grouping CPAs: small business requires general services and more personalized handholding operations, national accounts calls for more sophisticated problem solving with less need for the personal touch. Finally, in all instances, any CPA firm that is national in coverage tends also to organize around geographic centers—the New York office, the Chicago office, etc.

The geographic basis for organization is the most common one among national firms, and then within any one location firms will choose between organizing around functional specializations or around clients. When a client itself is national or global, however, the firm often experiences the need to integrate the services offered to each client in all its various locales. Here the matrix concept can help make explicit the need to organize along both requirements simultaneously. Similarly, all of the specialists in any one function need to keep their practices consistent, and they need to keep informed about recent developments. Managers of local units and account managers become the 2-bosses, and the shared subordinate may also relate formally to a technical expert for quality control purposes. (See Fig. 7.2.) In most CPA firms the coordinating activities along the secondary and tertiary dimensions are handled by use of committees and task forces, and the matrix need not evolve beyond what we called an overlay organization. Individual senior partners can and do serve as integrators, generally for the various specializations. We only know of one CPA firm that goes so far as to evolve dual support mechanisms, such as dual budgeting lines and dual personnel evaluations, but these are hardly drastic additions to the basic concept, and firms would probably benefit by introducing them, without significantly increasing the internal complexity of the firms' management.

In sum, CPA firms probably do not need mature matrix organization to be effectively managed, but the greater the type of services offered, the more likely it is that they should consider elements of the matrix management and design.

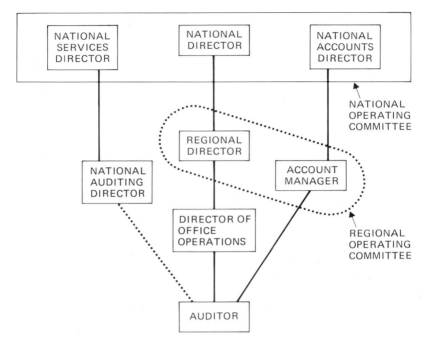

Fig. 7.2   Example of a CPA firm's organization.

LAW FIRMS

The most basic distinction in the organization of a law firm is between professionals and nonprofessionals. Professionals are subdivided into partners and associates, nonprofessionals are organized along typical staff lines and are supervised by an administrative coordinator and by one or more senior professionals. Specialization and departmentalization have been part of the large law firms' structure for a long time. Typically, these include litigation, taxes or estates, labor and corporate. Regional domestic offices tend to have the full range and organization of capabilities. Foreign offices tended to have competence only in tax and corporate affairs.

At the top of the organization, as a general rule, the larger the law firm, the more the ruling function tends to be delegated to an executive committee. Partly because of legal characteristics of a partnership and because of professional values, authority is shared and indirectly derived. The most distinctive feature of legal organizations, for our purposes, is that the values of professional autonomy prevent the formation of pyramids of power. Law firms, unlike corporations, have tended to be underbureaucratized. Few written rules exist, except for matters pertaining to clerical details such as docketing. The partnership agreement itself, for example, usually covers only interpretations of the state statutes governing formation and dissolution, and does not attempt to define the rights and obligations of members of the firm. Major committee decisions are made more by group consensus than by majority rule.

> . . . the managing partner ranks as the most powerful member of the firm, but the collegial organization of the lawyers places strict formal and informal limitations on his power over his fellow professionals, especially his partners.[1]

In this professional environment with a minimum of organization structure, the team concept is an important integrating tool between foreign and domestic branch offices, particularly when they are servicing multinational clients. Coordination does not occur through the superior authority of an individual. At middle and lower levels, project groups or teams are used, as in industry, to coordinate across both regions and specializations. Still, the tasks performed do not require a highly integrated structure; and common professional ethics, social and educational backgrounds, and client pressures provide both sources and limitations on "consensus and conformity." The result is that even when law firms expand their services offered and their territory covered, the growth is *not* followed by a subsequent increase in the complexity and formality of the firms' organization structure, management, or control. The matrix, as an alternative to the formal hierarchies that law firms eschew, can provide a structure that is at once more open and yet more coherent than the underorganized tendencies and managerial inefficiencies among large practices.

SECURITIES FIRMS

The expansion of the American securities industry reveals similar patterns of change in strategy and organization. The industry is extremely cyclical, and fortunes rise and fall with fluctuations of the Dow Jones averages. The bear market of the last few years has produced a high failure rate, and many old, large, and respected names have disappeared. Following industry life cycle patterns, the surviving firms have grown by diversifying their services and expanding their geographic coverage. As a result, they employ more people in more offices, and we witness a change in their organizational complexity. But, as with law firms, the changing task in terms of services and distances far exceeds the modest adaptations in their organization. Complex strategic change was followed by minor structural responses.

During the past five years most of the larger companies changed from partnerships to corporations, then they went public, and recently they have begun to form holding companies to allow them to enter and to coordinate better their diverse activities and subsidiaries. The diversification of risk to reduce highly cyclical profit fluctuations has led the industry to look to other economies with pools of money but with different economic conditions, particularly in Europe, Japan, and the Middle East. Sophistication in the sources and uses of funds has also increased significantly in recent years. Many members of the industry recognize the rapid internationalization of capital markets, and of securities markets in particular. Besides the well-publicized Eurodollar and petrodollar opportunities, American securities are belatedly realizing the interchangeability of currencies, businesses, and personnel. E.F. Hutton's global services, for example, now include: stocks; commodities; options; corporate, municipal, and government bonds; tax shelters; money market instruments; and personal financial management. Other services can include insurance, research, mutual funds, merchant banking, and underwriting of both foreign and domestic corporations in a wide range of financial markets and currencies.

Increased activity abroad generally leads us to a geographically organized international division. The division is usually wholly owned, often separately incorporated, and almost always located in their own separate headquarters—Paris and Geneva are popular, while Hutton for some reason chose Boston. Authority proceeds along geographic lines, both domestically and internationally, although some functions (services) are centralized and available to (or pushed upon) all divisions. Experts in most services are available in any major office, but their direct superior is the local office head, not a central specialist.

Many of the financial service firms that failed during the past few years did so because they could not handle, with sufficient speed and efficiency, the backroom chores. Speed and efficiency, in both order execution and in negotiation of deals, are all the more vital and difficult when operating on a global basis. A rapid and omnipresent communications network therefore becomes *the* critical basis for organizing a "full-range multimarket financial service organization" (Paine, Webber). The division of labor according to specialized services occurs on a geographic basis through the organization's structure; but the integration of these services and offices occurs primarily through technology and not through reporting lines.

Traditional methods of mail, telephones, and telex have been supplanted by round-the-clock private communications systems; computers make expertise in any branch office immediately available to any other office of the firm in the world. Goldman Sachs brokers, for example, use closed circuit television and satellite transmissions; and Merrill Lynch reports that communications, which is generally a fixed cost, has become the second largest expense, after labor, in the industry.

While the geographic and service diversification strategies of American brokerage firms has increased significantly in its complexity, the organizations' subunits have remained rather autonomous. "At least for now," according to one ex-broker, "most headquarters appear to concentrate more on coordination than control of their branches." Multinational growth in the brokerage industry will continue through firms seeking full-service expansion, by finding specialized market niches, and/or by their forming joint ventures with foreign brokerage houses, as in the British, French, American conglomerate of Warburg, Paribus, Becker Inc. Whichever pattern is followed, it is still likely that the managerial tasks and organizational forms of these large financial service firms will remain rather simple in contrast with the complexity of their task and with the complexity of management and organization in major industrial corporations. As with law firms, securities firms do not have need for the traditional and formal structures devised by industry. Where the legal field relies on professional norms of behavior to compensate for minimal organization, brokers rely on information-processing technology in lieu of formal reporting. The security firms that are diversified by both geography and services will nevertheless begin to experience the need for more coordinated organization, and the matrix offers them an organizing vehicle without a significant increase in bureaucratic structure.

BANKING

During the past decade the banking environment has changed significantly and, as a result, we have seen a lot of turmoil in the management and organization of banks. Many banks have broadened their activities far beyond their traditional charter; their new undertakings are often highly interdependent, and take place in complex and uncertain global environments; and it seems clear that only the big are surviving. These are the conditions that we have said are propitious and appropriate for use of the matrix. Despite this, it seems that only one bank—Citibank—has consciously used this approach. The case study at the end of Chapter 8 offers a detailed account of how the matrix works in the international part of the bank, but a briefer and more general picture is called for here.

The bank's management structure and processes reflect its specific objectives and global strategy: "to develop as a global financial service organization to achieve levels of competence, earnings, and size which compare favorably with the major corporations of the world." Accordingly, it evolved from a commercial banking base to a financial services organization that is diversified by geography, customers, services, and risk elements. The diversity is reflected in the dimensions along which it organizes: geographic, customer, product/service, and resource.

Geography.   Citibank has a presence in 103 countries, each with its own milieu of business opportunities and risks, competitive and regulatory pressures, and mounting social needs. Domestically, responsibility is shared along the other dimensions listed below. Abroad, there is an organizational unit for each country, with either a corporate or international officer in charge.

Customers.   The bank's key customers include corporations, financial institutions, governments, and individuals. An organization struc-

tured strictly along geographic lines clearly cannot effectively anticipate and respond to the special financial needs of each customer
group. Accordingly, the bank organizes around significant customer
groupings. For example, the World Corporation Group serves multinational corporation customers; the National Banking Group is
responsible for United States account relationships with domestic corporations not in the global classification. The Consumer Services
Group has responsibility for the total consumer market on a worldwide basis.

Product/Services.   In addition to traditional bank credit, the products and services offered by the bank include: leasing, factoring and
other asset-base financing, merchant banking, consumer finance and
insurance, management consulting, financial transactions processing, and fiduciary services. Many of these cut across geographic and
customer lines. For this reason, and from consideration of market
potentials, Citibank has identified certain institutional products and
assigned product management responsibilities including: market research, development and testing, and product profitability. Institutional investment services such as pension fund management and advi-
sory services, for example, are the responsibility of their Investment
Management Group; Traveler's Checks are managed by the Consumer
Services Group.

Resources.   Aside from the important human and financial resources
of the bank, the major resource that is in a matrix relationship with
the other organizing dimensions of the bank is its United States production base in the Services Management Group. This includes check
and other bank office processing operations, as well as financial transactions networks such as those for Citicard and for money transfers.

The responsibilities of units within each of these dimensions combine with those of others in the matrix. The optimal allocation of
funds in a particular country, for example, often requires a consensus
between segments of the geographic, customer, and product dimensions. The issues concerned in this particular decision-making process
are short-term versus long-term, fixed rate versus variable rate, local

versus global relationships, as well as interest rate differentials. Another example involves effective product marketing and cross-selling which require agreement on new business targets, introductions and revenues between the customer and product dimensions. To illustrate, marketing of the corporate trade and securities services of the Services Management Group relies on the active participation of the national, International, and World Corporation Groups. A third example, how the customer groups service their clients abroad by matrixing with the International Group in each country, is closely examined in the case study at the end of Chapter 8.

So far, other banks have not followed the Citibank lead in adopting the matrix. As the pacesetter in its field, it would not be surprising if others eventually followed suit; nor would it be surprising if, by that time, Citibank had evolved into something beyond the matrix.

RETAILING

To the best of our knowledge, the matrix structure has had only limited application to large-scale retailing organizations but we do believe it has considerable potential in this area of business. The clue that signals the potential is the existence of the long-standing tug-of-war that seems to go on in many large retail firms between the merchandisers and the operators. This struggle occurs in department stores, in supermarket chains, in discount houses, and in chains of specialty stores. Each side has had its days of dominance only to see the pendulum of influence swing back the other way. The operators are the guardians of cost, the merchandisers of quality. Both vie for control of the sales floor. Each side seems convinced of its claim that it makes the cash register ring. Firms in this industry have seen the tide of control swing back and forth several times. Is there no better answer? Perhaps not, but a matrix structure is being tried by a few firms.

The matrix structure as applied to retailing is quite straightforward. The sales unit manager reports to two bosses, the relevant merchandiser and the relevant operations manager. This arrangement is made explicit and ground rules are set up as to the areas of responsibility and direction that relate to each of these two bosses. For instance, one possible division of responsibility could run as follows:

Primary operations responsibility	*Joint responsibility*	*Primary merchandise responsibility*
Manpower scheduling	Inventory control	Pricing
Cash control	Contribution to overhead	Merchandise selection
Customer relations	Personnel appraisal	Merchandise knowledge
Floor appearance	Discipline and promotions	Product training
Store indoctrination	Compensation	Sales promotions and display
Fixtures		

Once some such system is in place in standard operational procedures and, what is really important, in people's minds, then a lot of the unnecessary hassle can go out of the retailing job. There will still be lively debates but no cold war. Neither side can win so they must learn to accommodate each other. As the economic cycle shifts, slightly greater weight can be given to one chain of command or the other without going overboard. A number of large retail organizations would seem to meet out Chapter 2 criterion of dual critical pressures. They may or may not exist in an environment that changes fast enough to generate high information-processing requirements or severe performance pressures. But, if these latter two conditions also hold, the matrix should be given serious consideration.

CONSTRUCTION—REAL ESTATE DEVELOPMENT

Firms that could be classified under the heading of construction and real estate development vary enormously in their size and scope of operations and the matrix approach could logically be an alternative option to only a limited number of them. Certainly a small general contractor need not worry about the fine points of matrix design. But the idea is proving to be of interest to a number of firms that are dealing with a wider range of the business opportunities in the field. We have in mind firms that have undertaken a number of large and varied projects and that provide a range of services such as site selection, design, turnkey construction, and real estate management. To such firms the matrix idea offers the possibility of arranging a set of specialized resources in their functional groupings in one arm of the matrix while the other would consist primarily of managers of active major projects. Many firms in this industry have evolved toward some such structure without explicitly formalizing the arrangement or using the term matrix. At the end of this chapter is a case study of the Carlson Company, a growing firm in the field, that has given explicit attention to developing a matrix organization. While the case describes many interesting aspects of the business, we would highlight here their handling of three related issues that are critical in this industry; project bidding, cost control, and economic cycles.

The Carlson Company has developed a structure and closely related organizational procedures that are especially well related to the economic uncertainties of their business. They have set up a number of fully owned subsidiaries that are each the equivalent of a functional department in a manufacturing firm and are each the "home base" for the varied specialists needed to execute all phases of a major building project. The heads of these firms act as the chief salesforce for their various services and often head up the bidding teams that direct specialists from several groups to put together sophisticated proposals. As the project proceeds the elected project manager is drawn into the team in anticipation of securing a contract. This ensures an orderly transition to the project management phase. The project office is given first-line responsibility for control of costs, schedules and quality with the backup of regular project reviews by the top-management team. The first has used the matrix to advantage in weathering major shifts in the availability of business by market segments, i.e., schools versus hospitals, and the highs and lows in the level of construction activity. They maintain a cadre of professional specialists and project managers that can be kept busy during the lows of the cycle and rapidly expanded by subcontracting for temporary services during the highs.

HOSPITALS

Hospitals today face a host of insistent pressures that are driving them through a jumbled set of difficult changes. Without dwelling on this list, we would cite such obvious issues as the rising expectations about medical care, the high cost of new medical technologies, the maldistribution and/or overcapacity of hospital facilities, malpractice suits, the mismatch of supply and demand for various medical professionals, uneven quality control, abuses of third-party payment systems, the potential of the computer, constant budget problems, and the continuing uncertainty about universal health insurance. In the midst of all these change pressures, one might well question any hospital administrator's sanity who spent time considering the complexities of matrix structures. And yet, it seems to be these very pressures that are pushing a number of hospitals down the evolutionary road toward a matrix. Not many of them are using the term. Not many are benefiting from the experience industry has acquired with this approach. Very few, if any, of them have evolved a mature matrix. But regardless, the new pressures on a number of hospitals seem to be creating the three conditions we noted in Chapter 2 that call for a matrix. Hospitals have always faced the pressure from the medical profession to find ways to utilize the advances in medical science. This process has been slow in some spots but the pressure has been real and the dominant position of medical chiefs of service in the hospital structure has reflected this emphasis in organizational terms. The newer pressures are coming from the "marketplace," the patients, and their spokesmen in the form of government and third-party payers. These pressures are mostly for getting control of rapidly escalating medical costs. Doctors and hospitals are not accustomed to responding to such pressures and are only now learning how to organize to become more cost-benefit effective. The overcapacity in hospital facilities in many areas is adding a note of competition, often for the first time, to enhance the pressure for performance.

The introduction of matrix thinking onto the hospital scene is not so much in the interest of loosening an overly rigid structure as it is to introduce more regularity into a fairly chaotic one. People working in hospitals do not have to be introduced to the idea of taking instructions from more than one superior. They are more apt to complain of having to take orders from everyone that wears a white coat or a surgeon's green coat. To classify relationships to the point where they have explicitly defined responsibilities to only two bosses would be a major step forward in employee relations and in cost control. Matrix thinking might end the fruitless debate as to whether the head nurse or the chief of service is the "real" boss of nurses and start a more rational dialogue about their respective and complementary authority and responsibility. In hospitals that are affiliated with medical schools, the responsibilities of chiefs of service would run not only to their academic department chairmen but also, in a defined way, to the hospital administrators, either lay or medical. Without delving further into the complexities of hospital organizations, let us add only that the major constraint on the more rational use of appropriate organizational structure is one of education: our medical system continues to place doctors with a fine medical training but with a total lack in management training in charge of some of the most complex organizations in our society.

HIGHER EDUCATION

American colleges and universities tend to be governed by tradition and have been little touched by systematic approaches to the management of education. Perhaps in the past this has been to the good since most management methodology was developed for achieving efficiency in mass manufacturing situations. We trust by now the authors have convinced the readers, if they needed convincing, that sophisticated management methods take full account of the differences between mass production tasks and R&D tasks. The stopwatch is no longer a useful symbol for modern management. This contingency idea carried over to higher education would sharpen our awareness that managing a graduate school of arts and sciences is not the same thing as managing a professional school, a liberal arts college, a two-year community college, a technical institute, etc. Yet the management of all higher education in the United States has been heavily influenced by the pattern established by prestigious universities geared primarily for graduate education. The dominance of specialized graduate departments, the rewarding of faculty for research instead of teaching skills, the emphasis on publications, has spread by emulation into all areas of higher education. These traditions have only recently been subject to sharp questioning by state and city governments that are challenging the cost and the quality of teaching in these various schools. The "marketplace" is beginning to assert itself. Some of the schools in which this process is further along are putting in place a second command structure to complement the tradition of dominance of the academic departments. These newer organizational units carry various names but they are essentially program offices set up to manage the resources committed to specific educational and service programs. These offices usually start as simply a coordinative overlay but some have evolved into a full-fledged second chain of command in their relation to the faculty.

This process is probably most advanced in professional schools and the authors have had personal association with such a dual command matrix as it has been explicitly used at the Harvard Business School for several years. We will briefly draw on the experience at HBS as a convenient example of the potential of matrix organization in higher education. The matrix as used at HBS provides a way of managing faculty resources—it does not impact directly on support systems such as buildings and grounds. All faculty are assigned *both* to subject area groupings (academic departments) and to program groupings (masters program, doctoral program, executive program, research program, etc.). These groupings are headed up by members of the faculty who serve for a limited number of years as area chairpersons and program chairpersons. Faculty members have responsibilities to both of these leadership roles. This matrix structure is supported by an array of additional mechanisms such as a simple dual resource accounting system, a dual faculty evaluation system, and a dual set of forums for developing academic policies.

In our judgment this system works relatively well, but not all of our faculty colleagues would agree with this judgment. One reason for the very mixed opinion is that the system is a very demanding one for the faculty. They really feel the pressure to perform in terms of the quality of educational programs as well as in terms of professional excellence. Probably the most sticky issue in the operation of the matrix is the process of assigning faculty to the various competing programs. This is a complex and delicate process under any system, since the professional needs and personal aspirations of the individual must be weighed along with the various institutional requirements. HBS is still seeking ways to refine and streamline the process. But even in this tricky area, program and area chairpersons are learning to debate and resolve their inevitable conflicts without neglecting the concerns of the individual faculty member.

While matrix management has not as yet proceeded very far in the realm of higher education, we do believe it has real potential there.

FEDERAL AGENCIES

Does the matrix have a contribution to make in improving the effectiveness and efficiency of government operations? Our answer to this question must be more speculative than in the other not-for-profit segments for the simple reason that there have been fewer applications to date. The authors are aware only of some limited applications in the military, some experimental work in HEW, and some steps toward matrix in the Environmental Protection Agency. The EPA will be drawn on for an example of what form a matrix might in the future take in a major federal agency—not an example of what is actually in place at this time.

Before moving into this example, we should consider why matrix forms have not developed further in large federal agencies. After all, at least some of the many governmental agencies must face the three conditions we have pulled out as criteria for adopting a matrix. The lack of matrix development may be a clue for some that such agencies seldom face the third criterion, stiff performance pressures. There is undoubtedly some truth in this. Beyond this explanation, we would suggest that our government's system of checks and balances in the use of public resources is a major restraint on the use of matrix forms. The legislature authorizes expenditures in the executive branch by predetermined categories. Expenditures must be accounted for by these categories to ensure the proper and intended use of public funds. Elaborate procedures have grown up to control fraud and dishonesty. This authorization process and related control procedures make it awkward to arrange for the joint responsibility for resources that is an integral part of matrix management. And yet, few will deny that the overwhelming problem of governmental organization is not the dishonest use of resources but their waste. If the use of matrix management can significantly improve the cost–benefits of some agencies, perhaps we can find a way around the existing accountability procedures.

The EPA was formed by the consolidation of a number of smaller activities drawn primarily from the Departments of Commerce, Agriculture, and Interior. It has been funded by Congress in terms of the large categories of water, and air purification and a set of smaller programs in regard to solid waste disposal, pesticide control, and disposal or containment of atomic waste. It operates a research and technology center and a number of regional offices. It clearly must do both an educational and an enforcement job. It also operates a program for subsidizing local sewage treatment plants. Figure 7.3 indicates one possible way of organizing this array of activities. It is shown primarily to indicate some of the possible ways to combine product funding around goal categories (air, water, earth) with geographic and functional units. To make such a structure work, only a limited percentage of the funds supplied by Congress to each goal or category could be spent on headquarters personnel directly under the heads of these bureaus. The bulk of the funds would be used to purchase services from the other side of the matrix, the regional offices and the research and technology center. Therefore, these latter offices would have the direct supervision of the bulk of the EPA personnel. It would be expected that this balancing of the power of funds against the power of direct supervision would create a useful tension between an end-use orientation and professional input of local and technical knowledge. Congress could hold the air, water, and earth bureau chiefs responsible for results, and the chiefs in turn could push to get their money's worth from the other side of the matrix by contracting afresh for services. Again, this is but a speculative example of the kind of imaginative approach to the design of governmental agencies that can be done. It will have bugs in it but with further learning significant improvements would seem entirely possible.

UNICEF

UNICEF, the United Nations Children's Fund, has long been considered one of the best managed of the United Nation's agencies. UNICEF is heavily dependent on learning in the field. Problems vary by country, by local politics and politicians, by local economic conditions and social and cultural norms, and in accordance with diseases and definitions of health unique to an area, country, or region.

Faced with this task variability, UNICEF works with a function-geography type of matrix, but with a difference. The structural arrangements are based on the assumption that the agency's greatest knowledge on a particular subject (e.g., nutrition policies for children) can as easily reside in an "office" situated in Africa as in a location in Europe or in Asia. The matrix is one of "knowledge centers" and

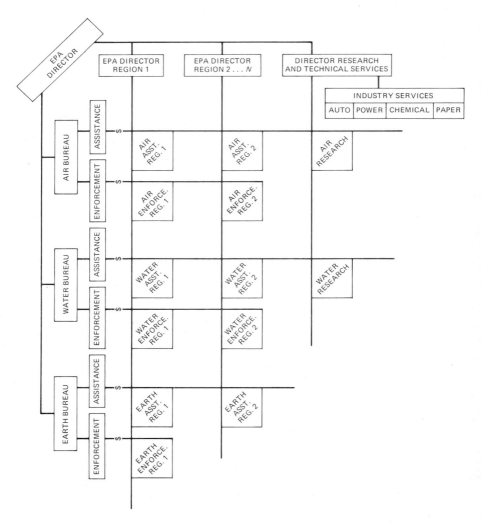

Fig. 7.3   Example for organization of the Environmental Protection Agency.

"offices" (regions or areas or countries). When special expertise develops in a given office, it is named the "knowledge center" for that particular specialty.

The person-in-the-middle in UNICEF'S organization is the program officer in a particular office. If his or her task is to concentrate on the nutrition problems of children in that particular office's geographic area of responsibility, he or she will report on the one hand to the UNICEF representative heading that office and to the leader of the knowledge center that has responsibility for the expertise relevant to the program in question. The program might require several areas of expertise and the program officer would, therefore, have to deal with several knowledge center leaders. He or she may even be the knowledge center for one of the areas of expertise. The program officer must have the flexibility to acquire resources from wherever needed to develop innovative solutions to problems. This can mean UNICEF knowledge centers or it can mean expertise drawn from universities, governments, etc. The matrix is uniquely suited to this type of flexibility. It legitimates and encourages the field members to look elsewhere than headquarters for the resources they require, a policy that encourages the learning of lateral relations. A diagnostic report from a consulting organization called in by UNICEF commented on this specific issue as follows:

> . . . we have seen in UNICEF a very constructive attitude towards risk taking and joint learning from problems, mistakes and failures, particularly in lateral contacts between peers. We have seen this in regional meetings and other gatherings. This is the major explanation of the innovativeness that UNICEF has shown over the years.[2]

HUMAN SERVICES -

> Problems can no longer be addressed through increased funding, rather they must be resolved through a more effective and efficient redistribution of existing resources. This does not necessarily imply that some people must go without services, rather vertical and horizontal consolidation must take place in areas of duplication and more effective and efficient interventions must be delivered to each client.

In order to achieve this, each agency at each geographic level (state, regional, area, city and town) must function as part of a larger human service system. Individual directors and staff at each level must begin building relationships of collaboration and mutual support so that they can collectively attack and solve these organizational problems. This is especially true in the operational area level where agency boundaries must be made consistent and a structure for achieving horizontal integration established.[3]

Working with a grant under the Title XX Act directed towards better integration of human services, the Department of Mental Health of the Office of Human Services of Massachusetts has launched an ambitious and innovative effort to rationalize the different human services available to its citizens. The project began in 1974 in the Taunton area of Massachusetts, a "catchment" area that contains 13 distinct towns or communities. The project is viewed as a prototype of how human services throughout the Commonwealth of Massachusetts might be organized. The final assessment of success will be a long time in coming, but the perception of success within the agencies of the Office of Human Services has been sufficient to attract several other catchment areas to structure themselves similarly.

Two organizational innovations underlie the project design: the use of "social networks"[4] which bring together the entire social system of a client (family, friends, relatives, employer, teachers, members of the clergy, etc.), and a matrix organization design. As with the federal EPA we previously examined, the duality of this matrix is function and geography. The functions are the traditional specializations contained within a department of mental health, often within a local hospital; for example, psychiatric services, social workers, geriatric specialists, counselors, physiotherapists, nutritionists, etc. The geographic focus centers on each of the towns or communities in the Taunton area.

The 2-boss managerial role in this matrix is the person responsible for the particular client who not only coordinates teams of the relevant specialists for the "case" in question but also organizes the support system within the social network. As with every person occupying this role, the 2-boss person must be at home in both orientations. She or her must be familiar with the type of resources available in each function, with the particular individual who would best fit the case or client or problem in question and she or he must have some subjective knowledge of the work load, disposition and capacity to work with (particular) others of each of the specialist resources available. On the other side, the 2-boss person must know the community intimately; the clergy who might help; the personnel who can advise about job availability at the local employment security office; sympathetic employers; local law-enforcement personnel; as well as those more directly connected with the particular client.

The Taunton area has viewed their organization as a "stacked matrix."[5] The Department of Mental Health-community duality is only one level in the matrix. The activities of the department must also be coordinated with other agencies of the Office of Human Services that operate in the same catchment area. Rehabilitation Service, the Department of Youth Services, Correction Services, Welfare, etc. all coalesce to affect a particular client's social network and, if not coordinated, can lead to overlap, redundancy, and unnecessary conflict. In a similar manner, the geographic area or catchment organization must recognize the regional administration and the specialist resources that exist at that level. Hence one matrix is superimposed on another and the concept of a "stacked" or layered matrix is developed.

An interesting by-product of the dual orientations that characterize Taunton's human service effort is the way the managers of the system have acquired a new type of learning. They now look at traditional dualities and view them as processes that can be integrated and need no longer be categorized as "either/or." In particular they make mention of the following dualities in their system:

- Categorical agency functions versus human service functions
- Decentralization versus centralization
- Community of interest versus geographic community
- Specialist versus generalist
- Internal cause versus external cause
- Short-time response versus long-time response

They see these items as capable of being integrated through "joint efforts," in particular, through: joint information intake; joint triage; joint referral; joint monitoring; joint follow-up; joint information systems; joint planning; joint personal development and joint budgets.[6] In essence, they've learned to think matrix in terms of integrating differences and to act matrix in terms of shared relationships. For the Taunton area human services, matrix has provided a way to achieve the "more effective and efficient distribution of existing resources" which they strived for.

NASA

In just about every new organization behavior textbook, the section dealing with organization design and/or structure usually has two or three pages devoted to matrix organizations. The diagram(s) used to illustrate the form are almost always derived from a NASA project, those projects or subprojects managed by or subcontracted from the National Aeronautics and Space Administration. The vertical axis of these diagrams illustrate the major functional groupings differentiated by technological specialty, e.g., electronics, power systems, mechanical, systems integration, etc. The horizontal axis illustrates the different project and/or program offices along with the various coordinative and administrative functions that the managers of the projects and/or programs perform. NASA's successes have aroused enormous interest in project management and matrix forms of organization and the extensive dissemination of NASA-related or -inspired publications (e.g., Department of Defense standards for project management) have gone far to publicize these forms of organization design.

In particular, NASA literature has emphasized the central role of the project or program manager. The functions of the program manager are to legitimate the program, obtain resources, plan activities, solve problems, coordinate functions and other resources, oversee informational recording, administrative and personnel services in the administrative and personnel services in the program, etc.[7] Since the list of activities is extensive, the task is frequently subdivided into several roles: program manager, program coordinator, many assistant program managers, program planner and several others. The total set of these roles in a NASA project is referred to as the "program office."

A feeling for the texture of these varied roles can be gleaned from the comments of two program office managers on the Meteorological Satellite Program at NASA.[8]

The program manager had better be a flexible and patient guy. He has little real authority but great responsibility because he has to be the middleman in negotiations. It is the program manager who has to arrive at the total objective and obtain funding by Congress. He is also the middleman who must bring together the various user groups, scientists, and experimenters in arriving at an agreement about the payload. . . . Persuasiveness, technical knowledge, personality, ability to sell—these are his tools.

A good program manager has to have a degree of aggressiveness. He is an individual who has to be comfortable working independently, knowing when to call for help, when to push things up through the system, when to bring in someone of higher status in order to resolve a difficulty that can't be dealt with at the functional level. There's no magic formula for handling most of these problems. It's a matter of negotiation and give and take.

Growing a matrix organization from scratch is a long process of learning for an organization. The only way to shorten that process is to hire new people who already have experience in many of the facets of matrix management. One of NASA's important legacies to the nation is the portfolio of trained program and matrix managers it has developed. Many of these have found employment in other organizations and have spurred the development of effective matrix organizations.

CASE STUDY: THE CARLSON GROUP

In February, 1975, almost 30 years after its birth as a small steel erecting company, the Carlson Group of Cochituate, Massachusetts, was a dynamic designer/builder with annual sales well over $30 million. The range of products it offered the "environment creation" industry was unique for a company its size. Separately or as a complete package, the Carlson Group performed all steps required in the process of creating commercial, industrial, or residential facilities. Beginning with planning the legal and financing vehicle, the company could accomplish architectural and engineering design, site preparation, total construction projects, subcontract specialties, and commercial or residential property managment. Jack Carlson, president of the Carlson Group, attributed much of the company's success to its organization structure, one unique in its area of business.

The Company's Markets

In 1974, total spending for construction had barely matched the prior year's level of $135 billion. Throughout the industry, sharply rising costs were squeezing earnings and fueling competitive pressures in the face of a decreasing physical volume of total construction. Material shortages and delivery delays were lengthening the time from start to completion of many projects, and new environmental considerations were imposing heretofore unknown constraints on architects and builders alike. The mix of construction was also shifting, with private residential building starts down sharply while private nonresidential building showed small gains.[9]

The president of the Carlson Group described the markets in which it did business as having historically been "cyclical, highly volatile, but overall good growth markets. Design/build growth has been especially dynamic." In 1974, with volume at $35 million, Carlson had been ranked the sixth largest builder in New England. As of January of that year, the company had approximately 15 percent of the $165 million manufacturing facilities market, 5 percent of the $150 million commercial market, 3 percent of the $125 million educational market, and 2 percent of the $190 million medical market, all geographically centered on suburban New England. It also controlled by land ownership approximately 10 percent of the $100 million residential building market in Eastern Massachusetts. Because of the diversity of services it offered, the Carlson Group had several different sets of competitors. It faced off against real estate developers, architects, engineering and design companies, construction management firms, and general contractors.

Company History

Carlson Construction Company began as a contractor for the erection of structural building steel, and expanded operations into general contracting. In 1965, under Jack Carlson, the firm decided to offer a concept somewhat unique in the construction business—design/build packages within the "total project responsibility" concept.

Design/build differs significantly from more traditional construction industry practice in which an owner who needs a new facility must deal with several different suppliers of services in the building process. Traditionally, an owner would first locate and purchase a site, then call in an architect to whom his or her needs would be explained. The architect would design the facility to the owner's satisfaction, then engage various construction contractors and subcontractors to have the building constructed. All of these contractors, however, would be looking out for their own best interests, and the architect (who also held self-interest paramount) would be expected to coordinate the efforts of the various contractors and suppliers. In this approach to facility design and construction, a continuing tug-of-war often ensued among the architect and all other firms involved, with all concerned attempting to maximize their own objectives. Frequently this worked to the detriment of the owner.

In 1956, with sales approaching $3 million, Carlson Construction Company offered an alternative. The concept was simple—Carlson provided a completed turnkey facility to the owner, who would have to deal with only Carlson from initiation to completion of the building project. Carlson assisted the client (owner) in designing the structure and locating a suitable site; with plans approved, it then supervised the entire construction and finishing process. Carlson had in-house capability in architecture, engineering design, and all phases of construction, so it could handle the project from beginning to end.

Advantages to the client were numerous. Carlson offered guarantees as to both maximum cost and maximum time to completion, with repayment to the client of a portion of Carlson's additional profits if the project was brought in satisfactorily under cost. Because all services were performed or supervised by Carlson, the traditional tug-of-war was eliminated, with further savings and decreased aggravation for the owner. And Carlson guaranteed its work would be completed to the client's satisfaction regardless of unforeseen difficulties.

The timeliness of the design/build introduction was proven as Carlson proceeded to push sales beyond $30 million by 1971. Subsequently, a modified organizational structure began to emerge. A decade of rapid growth in sales had provided commensurate growth to each functional area of Carlson's expanding competence (real estate, engineering, construction management, etc.). To facilitate marketing of the various competences, functionally specialized separate companies were gradually formed. In 1974, these specialized companies were legally combined to form the Carlson Group. The Carlson Corporation was now only one of the eight wholly owned companies which comprised the Carlson Group. Among the services offered by the subsidiary companies were real estate management; design/build contracts for medical, educational, industrial and commercial facilities; architecture and engineering design; construction management; planned residential communities; and certain specialties such as industrial bulk-storage facilities and mechanical systems design and construction.

Each of the subsidiary companies performed its services both as an independent entity and as a component of Carlson Group teams on total design/build projects. Each was a separate marketing or resource center with a president responsible for its growth, development, and maintenance. For example, the Aldrich Company was a planning, design, and engineering firm. It offered its services to clients both as part of design/build projects and as an independent contractor. Likewise, COMSCO (Construction Management Service Corporation) offered construction management services both independent of, and as components of, larger Carlson Group projects.

In mid-1974, Carlson had seen itself poised for another major growth spurt, with projected sales in two years of $60 million, and $100 million within four years being a realistic objective. The past three years had been spent consolidating the Group to prepare it for further growth and clarifying the organizational structure to improve its functioning. In the fall of 1974 the recession hit.

Organization of the Carlson Group

The presidents of the eight companies, along with Jack Carlson, Tom McCarthy (vice-president, real estate), and Tom Yenser (vice-president, controller) made up the Planning Committee. This body acted as top management for the Carlson Group. Jack Carlson had a nominally equal voice on the committee but retained veto power. The Planning Committee met weekly to implement the strategic goals Carlson set for the Group.

Stock in the Group was privately held—Jack Carlson owned 51 percent, Bob Carlson 20 percent, and John Carlson, Sr. 15 percent; the rest was held by other officers. A program to increase stock ownership by key employees distributed stock automatically in response to the achievement of annual corporate objectives.

The organizational structure of the Carlson Group was basically a matrix. One axis comprised the various technical and functional specialties; the other was represented by individual projects. The latter might be investment, design, engineering, construction, property management, etc., or any combination of these. Eight project managers were responsible for running the up to 40 projects underway at any given time.

Each technical/functional specialty was part of one or another of the Group's subsidiary companies. Each specialty had a designated manager who was responsible for growing, training, and maintaining that specialty as a resource to the Group. These "resource managers" allocated the efforts of their respective resource groups to projects on the basis of priorities set by the Planning Committee. Allocation of re-sources was extremely flexible; they could by redeployed quickly when the need arose. People assigned to projects were carefully selected to form effective teams able to achieve the objectives of the client, the subsidiary company, and the Group, without conflict.

Individual projects were supervised by marketing managers prior to contract award and by project managers afterward. Effort was made to select a team in the early stages of proposal which would stay with the project throughtout its life. This had proven expedient in estab-lishing working relationships with clients, as well as eliciting early commitment from the implementation team during developmental stages of a project. To provide continuity, the marketing manager at-tempted to involve the potential project manager, project engineer, and superintendent for construction just prior to final proposal to the client whenever possible.

All project managers were responsible for coordinating and over-seeing the contributions to their projects from the various resource groups. Because project managers were constantly in competition with one another for scarce resources, the interpersonal and political ability of each manager could have major effect on the progress of that individual's projects.

The Matrix in Operation

A project at Carlson begins with a call from a company who has heard about Carlson's design/build capabilities by word of mouth. They ex-plore Carlson's interest in making a proposal for a turnkey facility. A proposal team is quickly formed, consisting of a marketer (a "getter" in Carlson parlance), a planner, an engineer, and a representative of the Group at large, who might be any one of the members of the Plan-ning Committee.

Working with client representatives, Carlson prepares a proposal which sets forth the facility's location, design, construction schedule, and cost estimates. Up to this point, Carlson has spent $2000 to $20,000 on the proposal. If the proposal is accepted, proposal costs become part of the overall contract. If the proposal is not accepted, the potential client is responsible only for the cost of the outside con-sulting reports (soil samples, test drillings, etc.) which then become the client's property.

If Carlson determines it wants to do the job, the formal proposal is made. So far the proposal efforts have been directed and coordi-nated by the "getter." If the proposal is accepted by the client, a proj-ect manager is committed by the Manager of Project Managers, Roger

Mackay. Mackay bases his choice of appointee upon such factors as the client's location and market, "chemistry" of the team and client, and management expertise. When possible, Mackay prefers that the "getting" team recommend the project manager. However, he holds veto power over their recommendation. The Planning Committee can also veto any individual project team member. When a proposal is accepted, accountability for the project passes from the getter to the project manager. Unless Mackay makes a last minute change in his choice, the newly appointed project manager has already acquired considerable familiarity with the project.

The project manager now takes control, working closely with engineering, purchasing, and construction. The planner and getter remain on the team as consultants. The project manager is responsible for bringing each of the Carlson Group's resources to bear on the project at the appropriate time and in proper amounts. Here the matrix is again operative, as resources are allocated to the project by the various resource managers as required, then redeployed as soon as their part of the project is completed. Projects have a dual cost control system as they proceed, with costs being collected by both the project manager and a manager of Project Cost Controls. In addition to the direction provided by its project manager, a corporate overview of the project is maintained in the person of the Planning Committee member who served as part of the original proposal team. This person has no active role in project management; however, this person becomes active only by exception.

After the Group had committed itself to accomplish a project, the project manager assigned to the job has to tap the specialty resource centers within Carlson and apply their talents to the project in timely fashion. Inevitably, several project managers competed for the same scarce resources at the same time. In such situations project managers negotiated informally with each other and with the resource managers to determine resource allocation. Project managers recognized the priorities set by the Planning Committee when arranging compromises.

When resource allocation controversies could not be worked out at the project manager level, they were kicked up to the Manager of Project Managers, Roger Mackay, where they were almost always resolved. In the unlikely event they weren't dispatched by Mackay, they went to the Planning Committee. If the disagreement could not be reconciled at this level, a final resolution was made by Jack Carlson. Said the president:

> There is almost always agreement by consensus once the logic of a particular requirement is understood by the conflicting managers. The

absence of intracompany profit centers, combined with the fact that part of an individual's income is dependent upon Group results, accounts for the elimination of much conflict regarding allocation of people and other resources.

Control Systems

A two-level system gave Jack Carlson and the Planning Committee effective control over the Group's operations. Overall Group financial control was the responsibility of the vice-president controller, whose office was the focus to which control information was directed. Projects were treated as cost centers. Project managers were accountable for project cost results once the original estimates generated by the planners were accepted. Planners were responsible to the corporate vice-president of Cost Estimating. Control information was regularly monitored by Jack Carlson, who received copies of all project and corporate financial reports and was briefed weekly on the Group's progress by the Planning Committee. He also maintained a far-reaching informal control system, which he used to manage the day-to-day operations of the entire Carlson Group on an exception basis.

Organizational Coordinating Mechanisms

Several devices in Carlson's organization functioned to tie together the distinct entities which made up the total Group. Some were structural. These included the corporate Planning Committee; an Operational Planning Group which overviewed all ongoing projects; mandatory Project Review Meetings to oversee specific projects; and "linking pin" managers who occupied positions in several disciplines simultaneously. Roger Mackay was such a manager, with responsibility for all functional engineering disciplines on one hand, and all project managers on the other. All structural coordinating mechanisms were cross-disciplinary.

Other coordination was achieved in a spatial dimension, with the layout of offices and work space purposefully designed to cause interaction between individuals working in different disciplines. Still further coordination was provided by the company's philosophy, interwoven into its organizational framework; this philosophy highly valued mutual respect, cooperation and collaboration. Employees could be reprimanded for being uncooperative or unnecessarily confronting. According to Jack Carlson, employees' success in the organization was highly dependent on their ability to live by these values and informal rules. The entire appraisal and compensation system was based on the Group philosophy.

The Group's President

The personality and guiding touch of Jack Carlson seemed to permeate the entire Group. It was evident that much of the company's character had been purposefully created by its president. His management presence was low key, with almost total avoidance of authoritarian style. He talked easily with employees at all levels, with whom all conversations were on a bilateral first-name basis. He attended to the individuality of each person, respecting differences in work styles and exercising care not to impose his style or methods on others. He treated everyone with a respect which set the tone for the Group.

Mr. Carlson's executive responsibilities were shared with the Planning Committee. "I see my organizational role as twofold, really—I have direct personal involvement in strategic planning and frontier areas of the Group's operations, and I also monitor its daily operations at arm's length." After being intimately involved in the kickoff of new programs, including team-building efforts where necessary, Jack Carlson would recede into the background to let the operationally responsible people carry them out. He was available for advice and consent, but didn't interfere with ongoing programs unless requested to do so by participants. Carlson said:

> The Group is management- and knowledge-intensive, as opposed to being capital-intensive. Human costs are about two-thirds of our total costs. Two of the largest problems I face as president are: how to continue to motivate people over the lifetime of a career in a society with changing priorities; and how to create continuous growth for the company, which equates with opportunity for individuals, in the face of constant economic, political, environmental, and cultural change.

By mid-February, 1975, the more abstract concerns of the preceding paragraph had been displaced in the president's thinking by the massive immediate problems posed by the deepening recession. Said Carlson:

> Today's problem is, how do we react resourcefully enough to absorb the beatings of the economy? The industrial market is drying up in New England, and we're regionally based. The commerical market, largely suburban, is in crisis—here we're talking about large retail organizations. Many large retailers are in serious trouble; some have gone bankrupt. Industrial and commercial markets are traditionally two of our largest business areas. The real estate development business is in a crisis of potentially catastrophic proportions. Many banks have a moratorium on real estate loans.

So we're left with medical, educational, and housing. Housing is the most depressed it's been in 25 years, but we can convert this problem to an opportunity. Because of today's crisis, there are going to be fewer developers of housing in the future—that market is shaking out. We see housing very bullishly as it turns around. We own two major locations in the Boston area, both ripe for development—and we'll be facing fewer competitors. We're also better attuned to changing life-styles and environmental concerns than the competition.

As a result of the real estate development crisis, those organizations which are financially sound are being called upon by the institutional investment community to perform services for a fee to help them out of their troubles. This is a new market opportunity for us we couldn't have foreseen as little as six months ago, but we're willing and able to take advantage of it. The flexibility of our matrix allows us to react rapidly to new opportunities. For example, we've grown our business in the medical market from zero to $25 million in just 24 months in response to new opportunities there. We should be able to continue forward in both educational and medical markets. We can get more income out of the same amount of opportunity now, because of economic conditions which allow us to render new and valuable services.

The cutback we made in December 1974 was necessary for us to regroup, to get lean enough to survive and be able to go forward. We've retained adaptive people. We've therefore increased the flexibility and "hunting strength" of our organization. Economic indicators in New England tell us we have to reach out into other parts of the country.

As a result of these conditions, we have some fear and uncertainty in the company. This breeds greater tension in our personal relationships—and they're *elemental* to the successful operation of our matrix. It now takes longer to get things done, because people tend to question each other more. It has a wearing, fatiguing effect. Our personal relationships are being pressed hard.

Some of our specialized professionals also face severe pressures. For example, because of the energy crisis, economic conditions, and environmental concerns, I feel "suburban sprawl" is dead. So we must face urbanization of the suburbs, and this requires new design technology, different from what we're used to working with. Parts of our technology are thus becoming obsolete, causing insecurity among some of our professionals who feel threatened. It's even possible that invisible sabotage could take place, for example, by someone shooting down an innovative idea he found scary.

We try to alleviate these pressures on our managers and specialists. We strengthen confidence whenever we can, backing it up with facts and figures. We share information more often than ordinarily. I'm now starting to meet informally with people at all levels of the organization, without their managers present. Just me and them, talking. I hope to surface their questions and concerns, and answer them directly.

Interviews with Carlson Employees

During the fall of 1974, several professional employees were asked to talk about the company. They responded to questions about what it was like to work within the Group's organizational context, and about the ability of the matrix organization to adapt to future growth.

Bob Carlson, senior vice-president of the Carlson Group, and president of Carlson Development Corp. and New England Erecting Corp. (29 years with Carlson) said:

> We call the matrix concept "teamwork." The system works as well as human nature lets it work. After consultation with others, everyone makes decisions better on his own, so it's hard for people to work in teams, but not impossible; and I feel it is working fairly well.
>
> New employees of Carlson Group are reluctant at first to ask for help in their job. But the successful man here knows when he needs help, and how to get it—who to call on. We've had real hotshots come in here who didn't make it because they didn't know how or when to ask for help, mainly because they didn't think they needed it. We've created an atmosphere in which asking for assistance is okay, and it's also okay to make a mistake. One of the biggest problems employers in the construction industry have is widespread lack of self-confidence on the part of their employees, frequently due to lack of formal education. Unwillingness to admit you don't know something is almost a character trait of people in this industry, and it's a real barrier to growth and development.

Joe Celi, vice-president, Corporate Marketing, Carlson Group; president, Carlson Corp. (16 years with Carlson):

> For the most part, I feel our matrix works reasonably well. The biggest problem has been communicating to people it is a matrix, so they can better understand some of the frustrations they're experiencing on a day-to-day basis. When people find themselves swimming around in a matrix organization, they can get confused and uptight. But I think we've done a pretty good job organizationally, because our people are quite open and relaxed.
>
> A matrix helps an individual better assess the value of his own contribution to the company, because he knows more about the totality. Responsibility and accountability get pushed down in the organization, so people can better feel how they impact on the company. This allows employees to get job satisfaction earlier and more directly than in a more traditional organization.
>
> Carlson's matrix concept is dictated by the nature of our business, not our size. We're really just barely big enough to support the matrix. If we were to double or triple in size, we might do a better job of running the matrix.

Roger Mackay, vice-president, Carlson Group; Manager of Project Managers; and president, Aldrich Company (15 years with Carlson):

I know of no alternative to the matrix, with 40-plus projects going on at any one time. It's hard to set priorities properly and get everything coordinated. It takes some pretty good politicking to get the resources on your project when you need them, instead of on someone else's.

Having good project managers is a major key to making it work; so is having good project teams. Problems really arise for resource managers who have to establish priorities for where they'll put their support. The more experienced, credible project managers are able to get better support for their projects, because they are better able to manage the support. A project manager has to be able to communicate his needs to resource managers. Resource managers must believe the requesting project manager is telling them the truth about his actual time constraints.

Jim Wakefield, vice-president, Civil Engineering, Aldrich Company (five years with Carlson).

When you take a company that two years ago was only commercial and industrial, and you go into several new fields almost simultaneously, you need flexibility. The matrix organization gives us that. The system is great—efficient, productive, and gives us a recycle capability. I love the diversity—as a specialist, you can still be working on fifteen different jobs a day.

I'm responsible for my own priorities. If three jobs come in, I determine in what order we're going to work on them. I talk with people associated with the jobs to determine which should come first. I have to evaluate the person who's making the request to know how urgent the job really is.

I never kick interface conflicts up to Roger Mackay. We can always resolve them ourselves. But it sure would be helpful to have some way to coordinate the variety of specialties within the Aldrich Company. We have many disciplines here and the only way they interface is on a project. I don't think that's right. We need some mechanism to coordinate the operations of all engineering and design functions as a group. In many cases, civil engineering could profitably coordinate with mechanical engineering, electrical engineering or design. We need some controlling point at which to do this.

Frank Pesa, project manager, vice-president; Carlson Corp. (11 years with Carlson).

Carlson's atmosphere has always been oriented to personal relationships, which is necessary with this form of organization. I look forward to coming to work each day. Nothing is very repetitive, there are always new demands, new challenges. Carlson has always been rewarding to producing employees. My personal growth has been a good example. I started out as a draftsman eleven years ago and now I'm a vice-president of the Carlson Corporation and project manager with room for further growth.

Ted Kohler, field supervisor (ten years with Carlson).

When I first started with Carlson, I didn't like the matrix because I couldn't always get the answers I needed. Now I know my way around and I like it better. I like being given the responsibility for the full job. You're not a puppet as a field supervisor at Carlson, which you are for most construction companies. I do think, though, that people at headquarters would be more effective if they had some field experience, which most don't.

NOTES

1. W. Delaney and A.H. Finegold, 1970. Wall Street lawyer in the provinces. *Administration Science Quarterly* **15**:(191): 193.

2. United Nations Economic and Social Council 1975. Management survey of UNICEF. E/ICEF/AB/L.147, 17 March p. 13

3. W. Robert Curtis and Mark Yessian, 1975. Integration of human services in Massachusetts. Working Draft, Dec. p. 1

4. W. Robert Curtis, 1974. Team problem solving in a social network. *Psychiatric Annals,* December.

5. W. Robert Curtis and Duncan Neuhauser, 1974. Providing specialized coordinated human service to communities: The organizational problem and a potential solution. Working Paper, Taunton Area, Massachusetts State Department.

6 . Curtis and Yessian, *op. cit.,* p. 16

7. Andre Delbecq and Alan Filley, 1974. *Program and project management in a matrix organization: a case study.* Bureau of Business Research, University of Wisconsin–Madison.

8. *Ibid.,* p. 82–83.

9. Standard and Poor's, 1974. *Building Industry Survey,* August.

8
THE MULTINATIONAL MATRIX

"For the majority of managers," says Peter Drucker, " the [matrix] structure is not of direct personal concern—though any manager in a multinational business will have to learn to understand it if he wants to function effectively himself." [1] Is there some organizational imperative towards a matrix form in multinational firms? Companies that never considered matrix management, even those that are quite negative about it, often find themselves evolving elements of matrix design as they grow internationally. Because increased use of the matrix often relates explicitly to foreign expansion, the domestic organization frequently accepts and utilizes elements of the matrix only in its dealings with international activities. Drucker, himself, considers the matrix "fiendishly difficult," so why are global corporations turning to it?

In Chapter 2 we examined three critical pressures that the corporation must experience in the external environment, and how these are translated into management behavior and organization structure in the firm, before the conditions are right for a matrix to evolve. These conditions may be seen in a global environment and corporation as well as in domestic ones.

(1) Domestic organizations that develop matrices have experienced pressures for survival from two parts of the environment, simultaneously and somewhat competitively. These were generally experienced as the pressure for technical or functional expertise and the need

for business- or product-oriented management. Neither technical nor business expertise was subordinated to the other, and instead a balance of power and a shared decision making were achieved. These translated structurally into a balance between functional and product dimensions in the organization.

In the multinational corporation, technical and/or business pressures are experienced as equivalent to the pressures from operating in several geographic areas. The boundaries of these different locations may be determined by distance, nature, legal, and/or cultural considerations. Whatever the basis, they are numerous nevertheless, and they cut across technical and business needs. Each country, for example, will manage several of the firm's business lines just as any one of the lines may operate in several countries. A multinational company therefore must balance the decision-making power of its product-oriented managers with those of its country-based executives.

(2) Operating in several countries, which is the most usual method of partitioning geography, means working in environments that differ greatly in economic, political, social, cultural, and legal terms. Each country has its own complex rules for importing a new machine or diverting tax payments, each has a different degree of political stability, some are tied to Common Market areas while others are not, and so on. This means a variety of external relations, with varying degrees of uncertainty, complexity, and interdependence, must be managed, in addition to the relative uncertainty, complexity, and interdependence in the domestic side of the firm.

We saw that these environmental conditions required an enriched capacity to process information in a business, and this too is the case multinationally. Whether and when will the pound be devalued again and should it be hedged against? Will the Argentine government stand or topple? Will the workers in the Italian plant lock up the managers to prevent layoffs? Will worker participation on the board of the German subsidiary tie management's hands excessively? These are intelligence matters; gathering, evaluating, and communicating information in a timely fashion to the relevant points in the organization.

(3) In a domestic organization using the matrix, technical specialists are grouped together to secure economies of scale. Specialists that relate directly to the external environment, such as legal and market research, may be located outside of the matrix; others such as the various engineering groups, may be arrayed along a principal axis of

the grid. The country affairs experts of international business represent a pool of scarce human resources that must also be shared to achieve the same economy. They may be in operations or not, but there are never enough at a low enough cost to permit replication across product groups within any one location.

The situation is even more complex in multinational firms that have multiproduct plants in different countries. Scale does not enable the firm to have a manager for each product line, and so the various lines might share the same manager in a given country.

Each of these environmental pressures

- two or more critical foci simultaneously
- uncertain, complex, and interdependent tasks
- search for economies of scale

triggers managerial and structural responses in an international business. As with a domestic organization, one of these conditions alone is not enough to move a company into seeing the matrix form as desirable. All three taken together, however, do create a sufficient climate for the global matrix. To understand how a matrix takes shape in the multinational corporation we must first understand the evolution of the firm's international business and its organization.

THE INTERNATIONAL DIVISION

The transition from a domestic organization to a worldwide organization involves a number of phases in which foreign activities begin as a minor and peripheral part of the firm and end up being so central as to change the geographical basis for organization to global parameters. At first, the nature of demand is not yet well understood, only a small number of firms are involved, and they rely heavily on R&D, skilled labor, and short production runs. All manufacture is domestic and exports are limited to developed countries with high GNPs. This phase, historically, was characteristic of the early 20th century, when United States firms found that their strategic shift to product diversification stimulated the growth of exports, and this frequently became an important source of revenue for the domestic companies. Organizationally, however, this seldom meant more than the creation of an export office.

In the growth phase, as the technology and the development of the new products come to be understood, methods of mass production are introduced, price elasticity increases, and price competition begins. There are a large number of firms manufacturing the same product line and production starts in other high GNP countries. The early investment in direct foreign manufacture generally has been in defensive response to this threat to export markets by local manufacture. The foreign subsidiaries created in the early phases of a domestic firm's move abroad are generally quite independent from the managerial and administrative control of the parent. This initial period of subsidiary autonomy, however, is rather shortlived. Sixty percent of the 170 companies studied in the Harvard Multinational Enterprise Project historically grouped their subsidiaries under an international division after the acquisition of only their fourth foreign unit.[2]

While the structure of the domestic company is laid out along product and/or functional lines, the international division is organized around geographical interests (Figure 8.1). The head of the international division is on a hierarchical par with the heads of the domestic product groups, and all report directly to the president. General managers of each foreign unit report up to the boss of the international division, and the units themselves reflect the same functional organization as exists in the domestic product divisions. With an increase in the number of foreign manufacturing units in any one geographic area, an intermediary level of regional direction (e.g., Vice-President Europe) is usually created between the subsidiaries of affiliates and the head of the international division. During the early period of its existence, the international division has little staff of its own, and what staff does exist frequently is more closely tied to its functional department at the corporate level than to the international area division.

The initial impetus for the creation of an international division is to congregate the activities whose specialized character is that they occur outside the borders of the home country. For the parent corporation, the locus of foreign expertise comes to reside here, and the coordination of functions and products is still largely internal to the international division. Coordination between the domestic and foreign sides of the enterprise are very loose and are not given much attention except at the top of the corporate hierarchy. From the corporate perspective, the formation of an international unit gives legitimacy to a

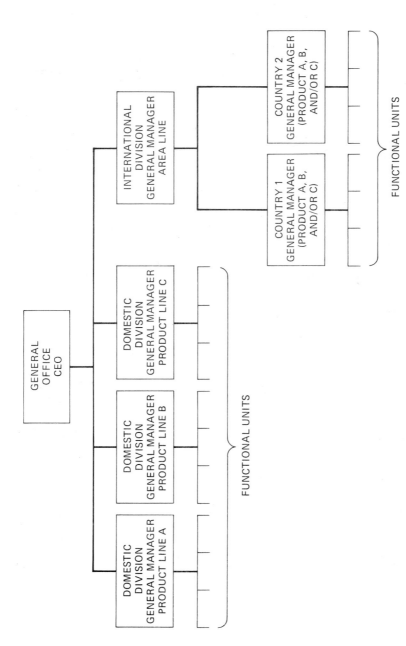

Fig. 8.1 Organization structured by domestic product divisions and international (area) divisions. The chart shows line positions only, not staff.

policy of multinational expansion, not formally recognized and explicitly stated before, and it provides an organizational base from which to develop. In its infancy, the division most generally turns its attention inward and develops under an umbrella of the benign neglect of the domestic company. For the foreign subsidiaries and affiliates, the creation of the international division provides guidance and support, but it also increases the control of the center over the periphery and reduces some of their previously enjoyed autonomy.

The creation of an international unit enables the firm to balance the self-interests of individual subsidiaries for the benefit of the company's total international performance. This can be done only by standardizing information and controlling some aspects of the subsidiaries' activities. Once this is done, taxes, for example, are minimized through the transfer price established for goods and services that move between sister subsidiaries in different countries. Central coordination of international activities also enables the company to make more secure and more economic decisions about where to purchase raw materials, where to locate new manufacture, and from where to supply world customers with product. Also, when the financial function of an international division is coordinated, investment decisions can be made on a global basis and overseas development can turn to international capital markets, instead of just local ones, for funds. The benefits of coordinating some activities on an international basis, rather than country to country, contribute to centralized control at the division level and above. But other activities, such as in the marketing function, often must remain specific to each local environment.

The success of the international division rationalizes the organization of a company's activities abroad, but it also creates a dual structure that ultimately works against the benefit of the corporation as a whole. "Even with superb coordination at the corporate level, global planning for individual products or product lines, at best, is carried out awkwardly by two 'semiautonomous' organizations—the domestic company and the international division."[3] The joint problems of specialization and coordination have been raised to the top of the corporate hierarchy but coordination also has to occur at lower levels in the organization and cannot all be bottlenecked in the president's office. Pressures for reorganization on an integrated worldwide basis become irresistible and the next question becomes, "What should that basis be?"

THE GLOBAL ALTERNATIVES

Disbanding the international division and adopting a global form involves making a basic choice in design:

- Is one organizing dimension going to be paramount and the others hierarchically sequenced in below it?

or

- Are two dimensions going to be considered equally important and balanced simultaneiously at the apex of the corporate design?

The first approach maximizes *either* the product, the functional, *or* the geographic dimension, while the second leads to a multinational matrix. Together they reflect four alternative forms.

The Global Product Design   In the global product design, the international division is carved up, and its products are fed back into the rest while domestic units become worldwide product groups. Products that require different technologies and that have dissimilar end-users are logically grouped into separate categories, and the transfer of products into various world markets is best managed within each distinctive product classification. The product diversification may be in related or unrelated lines. A strategy of global product diversification requires heavy investment in R&D; and the global product structure facilitates the transfer and control of technology and new products between domestic and foreign divisions. Firms that move toward a global product design tend to be in high technology industries and are often in OEM markets. Later in this chapter we will look in more detail at the Eaton Corporation that has adopted such a product format with the addition of a geographic overlay.

The Global Functional Design   The second alternative, a global functional design, has been adopted by a smaller number of firms. It tends to appear in basic industries such as aluminum and oil. These firms generally deal with raw materials, are vertically integrated, and have a single business line. The use of this approach in aluminum creates a worldwide division for the extraction and processing of bauxite, a worldwide division for smelting aluminum ingots, barstock, etc., and a worldwide division for fabrication and marketing of finished products. In oil the breakout would be along the lines of exploration and crude production, transport, refining, and marketing. This structure emphasizes a depth of process specialization regardless of end products, final customers, or world location. This global function approach is usually mixed with alternatives one and three.

The Global Geographic Design   The third option to create a global structure is based on geography. The domestic business is labeled as the North American area and the regional pieces in the international division are elevated to similar status. In contrast to the diversity and renewing growth phases of the product structured firms, companies that elect an area mold tend to have (1) a mature product line that (2) serves common end-user markets. They generally place great reliance on (3) lowering manufacturing costs by concentrating and specializing production using stable technology; on (4) marketing techniques as the competitive basis for price and product differentiation; and on (5) sensitivity to local regulations and tastes. Industries with these characteristics that have favored the area structure include food, beverage, container, pharmaceuticals, and cosmetics. The worldwide area structure is also highly suited to mature businesses with narrow product lines, because their growth potential is greater abroad than in the domestic market where the products and brands are in later phases of their life cycles. Since they derive a high proportion of their total sales from abroad,[4] intimate knowledge of local conditions, constraints, and preferences is essential. Many such firms rely heavily on advertising, and benefit from standardizing their marketing as well as production techniques worldwide. But standardization and area varigation are sometimes incompatible. In one classical gaffe, for example, advertisements for a major banking firm based in the United States used a picture of a squirrel hoarding nuts. The idea was to convey an image of thrift, preparedness, and security. When the same advertisement appeared in Caracas, however, it brought a derisive reaction, since Venezuela has neither winters nor squirrels as we know them. Instead, the image evoked a thieving and destructive rat. The major advantage of a worldwide area structure, then, is its ability to differentiate regional and local markets and determine variations in each appropriate market mix. Its disadvantages, as a form of organization, are its inability to coordinate different product lines, and their logistics of flow from source to markets, across areas.

The Multinational Matrix   All three of these design alternatives, however, represent trade-offs in which the one that is ultimately selected appears to have the greatest advantages. But what about the advantages that are lost by not having chosen the other design? The attempt to answer this question and achieve the advantages of several designs simultaneously has led many multinationals towards elements of matrix management and structure.

The multinational matrix represents basically a balance of power, information, and resources between the product and the geographic dimensions of the firm. The two dimensions are blended and given co-equal weight even though power balancing in a multinational matrix seems to be more difficult than in a product/function matrix. In such a global matrix, for example, the general manager of a French subsidiary will report to a vice-president for a worldwide product line as well as to a vice-president for Europe.

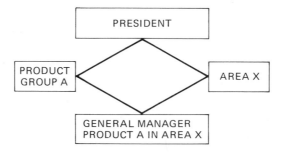

There are two separate occasions when the matrix is called upon as the preferable design, and the purpose of the matrix in each occasion is significantly different. The first is the use of the matrix as a transitional form to bridge the gap between the domestic and international sides of the corporation as the organization moves to some type of global structure. The second use of the matrix is as a more lasting way to deal with the three environmental conditions we established in Chapter 2. In addition, a number of firms that adopt a single command global structure, whether it is based on product, function, or area, find they need to coordinate around the neglected dimensions by applying some limited aspects of the matrix idea as an overlay structure. In the remainder of this chapter we will be examining in some detail the experience of four multinational firms that exemplify these various applications of the matrix idea and some of the problems associated with its use.

MATRIX AS TRANSITION

Based on hindsight, we can see in Corning's experience that they used matrix as only a transition between a domestic–international split structure and a global organization. This temporary use of the matrix, only as a transition, probably was not intended but others might follow this precedent by design, and, it is hoped, with less trauma.

Corning Glass Works is a billion-dollar company with products in four industrial groupings: consumer, science and technical, electric and electronics, and miscellaneous. International sales represent almost a third of their total, and almost half of this comes from its two largest foreign units. Half of foreign income comes from exports, and the rest from operations, sales of technology, investments and financing. Capital investment abroad, during the past decade, increased from 1.0 percent of the company's total to almost a third.

The company has long had a domestic bias, but the international organization grew nevertheless. While the parent corporation was organized by product divisions, the international company was set up geographically. Product coordination across these areas became a serious problem in the early 1970s, and the company created three international business (product) managers to complement the three area vice-presidents. The area managers were in charge of all Corning business in their part of the world, while the business managers were to be responsible for one business line across all geographic regions. All of these managers reported directly to the president of the international organization.

This first structural and managerial change was limited to the international side, and the domestic business was neither very interested nor threatened. The ability of the business managers to establish a worldwide product–market strategy was compromised from the outset, in terms of both structure and process. The dual reporting structure was intended to integrate the domestic–international split, but both reporting lines existed within international only and not at the corporate level where the bridge had to be built. The power conflict was present but never dealt with.

The new appointees had to rely on their personal skills especially because there was no *a priori* assumption on the part of the domestic division managers regarding the legitimacy of the dual structure or of these intended integrating managers. Each business manager took a different tack. The first tried to operate as a line manager, working directly with the foreign subsidiaries worldwide. With no established legitimacy, he ran into immediate trouble from which he never recovered. The second tried to outflank the area vice-president on the marketing front. An entrepreneurially oriented salesman, he tried to set up a worldwide trading entity for his product line and to reduce all foreign subsidiaries to blue-collar cost centers with no control over pricing. Only the third business manager tried to facilitate domestic–international coordination. He saw his role as an integrator, someone who facilitated global strategic planning.

One year after these jobs were created, a second organizational change was instituted in the name of creating a global posture. World Product Boards were established to help integrate each business on a worldwide basis. This was especially important in the sourcing and pricing of OEM goods sold to other multinational companies who were customers on a worldwide basis. Each board had about six members drawn from both domestic and international sides of the company. Senior corporate management committed themselves to the importance of the World Boards even more than to the Business Manager concept. Despite this apparent commitment, both concepts were abandoned in a major shake-up in mid-1975. The inevitable personnel shuffle took place and the international organization itself was drastically simplified. A large management consulting firm was involved in the reorganization at a cost of more than half a million dollars.

This third attempt to create a global organization is now being instituted by the consulting firm. It calls for worldwide product divisions, and it was initiated in a time of fiscal crisis. There has been a sense of imposition and submission. Again, the structural design is not inappropriate, but the costs and tensions experienced are from the insufficient attention to the process of change. Because the earlier steps were labeled "matrix," this form developed a bad name and the reaction has led to building the global organization around a more traditional hierarchy. Despite, or perhaps because of its unpopularity, it made people aware of the need to integrate the domestic and international parts of the business. It also served as a transitional design, not to be used in either the stable domestic period or with the new global product design.

A multinational matrix design can be used as a temporary solution in the changeover from the domestic/international division split to a global single command design, but whether it is viewed and experienced as the purveyor of good or bad tidings depends on how the change process is managed more than on how the matrix is structured.

One moral from the tale told above is that firms assuming a global strategy would do well to adopt a global structure. When they don't, it is likely to be because of history, politics, and personalities in and of the firm. Companies that want to avoid domestic–international splits among their management should avoid the same in their structure. When the matrix is called upon to heal the split, whether successful or not, it is a temporary form used to assist in the passage from a domestic firm with international leanings to a global corporation. Many firms make this strategic passage without ever needing to make transitional use of a matrix.

THE OVERLAY

Some multinational companies, while adopting a global single command structure, have found it highly useful to employ portions of the matrix idea in the form of a geographic overlay. Eaton represents a typical use of such an overlay arrangement in a single chain of command firm.

Eaton Corporation is a highly diversified company in the capital goods and automotive industries. It has sales of over $1.5 billion, employs over 50,000, and operates over 140 facilities in more than 20 countries. In 1974 each of its four worldwide product groups had a managing director for European operations. Each of the firm's 29 manufacturing facilities and six associate companies in Europe reported to one of these four people. In addition, 18 service operations, a finance operation, and a R&D center in Europe reported to their functional counterparts in the parent company. Senior management was concerned about how well it was coordinating all these activities in Europe.

It was important for Eaton to be able to evaluate and respond to significant trends and developments in European countries, such as tariffs, tax matters, duties, government legislation, currency fluctuations, environmental controls and energy conservation, co-determination and industrial democracy, labor matters, nationalization, and government participation in ownership. Its current organization structure did not provide a regular and convenient means of communication among its various European units, either for exchanging information, for building a positive corporate identity, or for assessing corporate needs and coordinating programs and procedures to meet them.

Rejecting the notion of country managers, and/or of one vice-president for Europe, instead they instituted a European Coordinating Committee (ECC) together with coordinating committees in each country where they had major involvement. The four European managing directors were permanent members of the ECC and each served as a coordinator for one or more of the country committees; Europeans representing various functions were appointed to one-year terms; and the firm's executive vice-presidents and group vice-presidents were all made ex officio members, with one being present at each ECC meeting on a rotating basis. Meetings are held monthly, midway between the monthly meetings of the corporate Operating Committee, and minutes must be sent to world headquarters within five working days. Attendance is required, the chair rotates periodically, and the location rotates around the major facilities. The president and the four group vice-presidents flew to Europe to formally launch the new coordinating committees, and the corporate newsletter devoted an entire issue to the new development.

Six months after the European Coordinating Committees began, Eaton formed a Latin American Coordinating Committee, and about a year after that they created a United States and Canadian Division Managers' Council using the same model. The same attention was paid to details of the committees' operations and to their implementation. The European committee, then, served as a model for realizing better coordination across business lines in each of the firm's major geographic concentrations, and it is not unlikely that the capstone in the future will be a council of councils that will take the form of annual or semiannual worldwide coordinating committee meetings.

Global corporations that are organized along product lines, such as Eaton, have to coordinate their diverse activities within each country where they operate. Having made the strategic choice to carry a diversity of products to new areas, their structures reflect the need to maximize technological linkages among the far-flung plants in each business unit. This has been done, however, at the cost of duplicating management and organization in each area. To cope with problems of coordinating and simplifying these parallel managements in each area, firms must reach through their existing product structure and weave an additional dimension across the organizational pattern.

The Eaton example is a moderate step, in structural terms, towards complementing the warp of a traditional product–line organization with the woof of geographic coordinates. The fabric of the organization is not significantly altered, rather it is reinforced. Little is done to increase the complexity of the global design or management practices. Success depends on thorough implementing of a plan that least disturbs the existing managerial style and corporate culture. The change is supplementary, rather than radical, and it has the desired effect of managing *both* product and country diversity.

INSTABILITY IN THE MULTINATIONAL MATRIX

Experience so far with the worldwide matrix has indicated that firms have some trouble in maintaining a power balance between area and product divisions more so than in the domestic product/function matrix we have primarily been examining. This is probably true because, by its very nature, geographic distance impedes the flow of the rich information carried by personal contacts and these unbalanced information flows can be reflected in power disparities. We can see some of this happening in the experience of Dow Chemical with a worldwide matrix.

Dow Chemical Company has used the matrix in its global structure for several years. Although Dow does not publish organization charts for internal or external consumption, its 1968 annual report nevertheless did publish a matrix diagram, of sorts, in the form of a photo cube. Along each dimension of the cube were photos of the key managers for the various functions, product groups, and geographic areas in the Dow organization. At that time the Dow organization philosophy was that they managed with a three-dimensional matrix.

While these ambitious notions of multidimensional structuring grappled with managing global complexities all at once, they proved exceedingly difficult to keep in balance. By 1970 it was apparent that Dow Chemical's matrix was effectively two-dimensional, a worldwide grid of product and geography with functions variously located at different levels in the grid hierarchies. In 1972 the matrix became further imbalanced when the product dimension lost line authority and was kicked upstairs in the form of three business group managers who reported to Corporate Product Development. They each were to be channels of communication for their product group across the areas, and their clout came from their control over capital expenditures. Life Sciences was the only product division that maintained worldwide reporting control.

Around 1974 Dow Chemical held a meeting of its senior managers worldwide. During an anonymous question-and-answer period with the chairman, Carl Gerstaker, the question was asked: "Which dimension of the matrix do you consider to be most important?" The very fact that the question was asked demonstrates that the matrix had deteriorated significantly. Gerstaker's answer was to the point: the most important dimension in a matrix organization is the weakest and/or the most threatened. Despite the chairman's understanding of multidimensional structures, however, the matrix continued to decompose. In 1975, the Life Sciences Division lost its worldwide reporting line and was subsumed under each of the geographic "operating units." Whereas each product used to have an identifiable team linking its business through the areas, the basic locus of these teams now exists within each area. Today, Dow Chemical would be described more appropriately as using a geographically based structure. In retrospect, it should be noted that, with the exception of Life Sciences, only the areas ever had their own letterhead stationery. Although the ideology of global matrix management still exists in some corners of Dow, the ethos and spirit of it is not to be found. They simply have found it too difficult to maintain a power balance.

AN EVOLVING WORLDWIDE MATRIX

A major corporation that has used the matrix successfully is Citibank, the second largest bank in the world. As discussed in Chapter 7, Citibank relies heavily on the matrix throughout many of its organizations. The case study at the end of this chapter details how its global matrix works, so we will only set the context and overview here.

In 1968 Citibank reorganized its domestic business from geographic lines to four market-oriented banking groups (personal, investment management, commercial, and corporate). A fifth, International Banking Group, maintained a geographic structure. In 1973 about 60 percent of the bank's total revenue and net income came from its international activities. It was then frequently said that the bank's senior officers abroad gave borrowing preference to local, rather than international, corporate clients in their belief that the returns were greater and that local clients did a greater volume of business in their country than did the foreign corporations. An improved data base demonstrated that the opposite was true, and the bank began to reorganize.

Multinational corporations had been serviced by the bank geographically. This meant that the domestic business of MNCs based in the United States was handled by the Corporate Banking Group, while the international business of these same firms was serviced by the International Banking Group (IBG). To eliminate this fissure, the bank created a World Corporation Group (WCG) to service multinational corporate clients around the globe. This organizational change took 223 major accounts away from the domestically oriented corporate group and took away the overseas business of these corporations from the international group. In addition, 234 multinational corporate clients based outside the United States were moved into the new group, again being taken away from the international and corporate groups. Differentiating this market segment on a global basis was strategically prescient, but integrating it into the existing structure of the bank and managing the internal political reprecussions was every bit as difficult and inventive.[5]

Each of the 457 multinational corporate clients has a Citibank team of ten to thirty people responsible for servicing that account. Each member of the team is stationed in a country in which that client requires significant banking services. Whereas the international group might have thousands of employees in a given country, the WCG might have only a few dozen. This means that the WCG must rely on the IBG organization in many ways, and the bank has been careful to spell out the interdependencies and how they are to be managed in detail. In all designated countries, for example, senior officers in the international group retain the overall direction and responsibility of total country market strategy, funding, and administration of corporate policy and practice. The head of the multinational corporate business in that country has to manage his affairs "within the country strategy guidelines established by the [senior international officer]." The senior international and corporate officers in a country are jointly accountable for, and evaluated on, World Corporation market results in their country. "While the WCG department will function as a discrete market management unit," says Chairman Wriston, "country organizations will provide MIS, operating, premises, and general administrative services and support." This kind of organization creates a managerial environment of high interdependencies between and within geographic and business market units. The results have been significant: profit center earnings in 1974 were 60 percent higher than 1973, with the same number of people, and 1975 earnings were about 50 percent above those in 1974.

The successful creation of the World Corporation Group in 1973, together with other major organization changes in such businesses as leasing and merchant banking, led the principal executive officers executive officers toward discussions of an evolving consistency in design. A coherent model emerged in October 1975 with the announcement of the creation of the Consumer Service Group (CSG). The CSG was given line responsibility for the consumer, or retail, banking business worldwide. The president's memorandum announcing the change to all officers began the section on organization with, "In keeping with our concept of driving core businesses through the IBG framework. . . ." The international side of the bank had been run as a line organization, and the senior officers in each country had been lords in their domain. Now, as expressed in a phrase of one executive vice-president, they are "integrators of sensitive processes."

With the matrixing of "core businesses through the IBG framework" the unity of command in each country has yielded to balancing of power based on country geography and business markets. Neither dimension can carry the organization by itself, and management has evolved a "concept" both for organizing and managing simultaneous, yet potentially competing, requirements.

From the first day of its existence, WCG was a significant business, representing about a quarter of the bank's total revenue and even more of its income. It also had a small number of employees. CSG, on the other hand, is in many markets an embryonic growth business that will expand significantly in the next few years, and yet it already has thousands of people working for it in each of a number of countries. Balancing this may be the fact that consumer banking doesn't threaten the existing businesses, and that those engaged in it represent a different culture within the bank. Still, it is an inescapable question whether, as the CSG burgeons worldwide, it will integrate well with the existing geographic framework or will come to rival it in parallel but competitive fashion.

If the previous hegemony is no longer a "business" itself, but is to become a "framework" instead—a mere scaffolding—then reactions are going to be quick and forceful. They came about two months later when the international group decided to recreate the same business–area matrix within its own unit. Lending money to governments, they said, was a different task from lending to other kinds of clients and should be managed separately; and the same for financial institution lending. These two activities therefore were taken away from the line responsibility of the country heads, set up as business lines within the power umbrella of the IBG, and matrixed with the country management. The titles of the senior international officers in major designated countries was changed to Area Corporate Officers, reflecting the sense that they belonged to, and must identify with, the parent institution and not just the international side of the bank.

The model that emerges is complex and sensible, but balancing the interest groups and blending them into a coherent institutional perspective will be a much more difficult task. The strength of Citibank's emerging global matrix is that it provides a unifying framework around which to organize diverse strategies and future innovations. The difficulty is that continued success with the model depends on the senior players' willingness to institutionalize shared powers. The organization is purposefully built around a paradox of competing claims; stability rests in managers' behavior more than in structural form.

A possible reason for the relatively healthy state of the matrix at Citibank at the present time is that there are at least three major competing cultures in the corporation: banking, international, and managerial. Each culture has its own management style, history, and patterns of behavior; each culture is embodied in an executive vice-president who is a potential candidate for replacing the president who will retire sometime in the next few years. Historically, then, the bank is in a period of openness and competitive exchange in the best sense. It is realizing much of the creative potential in such diversity. One candidate will either emerge or be selected, and it is during and after this process that a potential drift of the matrix towards a single dominant tone and line will be tested. Presently, however, Citibank is evolving a conceptual clarity in global corporate design, and it is institutionalizing styles of behavior to support management actions that are at once both diverse and cohesive. As such, it is becoming a model of advanced trends in organizing and managing a global corporation, for both industrial and service sectors.

The case study that follows takes you inside the bank and describes some personal reactions to the changes brought about by the matrix.

CASE STUDY: CITIBANK
Multinational Corporate Banking

January 1, 1974, the World Corporation Group (WCG) at Citibank began its first business day. The group had been created to deliver the full range of the bank's services to 457 multinational corporate clients. This case study describes the reactions of various affected managers after the WCG had been in operation for a year and a half. It concludes with a discussion of alternate models for structuring new businesses, and of the possible organization of the bank in the future.

Early Resistance

One of the senior vice-presidents in the old Corporate Banking Group was that group's liaison for enacting the reorganization decisions. Subsequently, he joined the new WCG. Although most of the bank officers spoke of the extensive predecision conversations to prepare people for what was to come, not everyone felt this way.

> When the bank announcement was made I can tell you it came as a big surprise for the entire organization. It came out of the Corporate Planning area on the fifteenth floor. About thirty-five key Senofs (the Senior Officers in each country where Citibank operated) were brought to New York, and together with the key officers they met at the University Club, at a now famous meeting. It was assumed that the Senofs would be resistant and upset because it was carving out, in a line sense, significant parts of their markets. It had to be done very carefully and very skillfully, and it was. Wriston started the meeting and went into the reasons for the change. The Senofs pretty much sat there and listened to him. Their reactions were as varied as the number of people there. Some of them said "It's been a long time in coming. It's a great idea." I was told on the side by other Senofs, "You guys are out of your minds. You can't make money against those customers; they don't constitute any substantive part of our market." There were some real misunderstandings about the true significance of this business.
>
> After this meeting, the processes of selecting firms (to move to the WCG) was fast, but very intense. Strangely enough, the real emotionalism came on the subject of whether or not it was a mistake to break up an in-

dustry group of companies. Could you have two petroleum departments, one within the National Banking Group (created from the old CBG) and one within the World Corporation Group? Or two mining groups? Even today they'll tell you, "Well, you know the mining industry is a very small one; they all do joint ventures together. The fact that some are multinational and some aren't really isn't significant, because the only minerals are where God put them," and all that razzmatazz, and "All mining engineers are first cousins."

There is no question that in 1953 we started to specialize by industry groups. The bank flourished tremendously against those particular markets because of the more intensive knowledge that we developed about the businesses. People who had been in those departments for any length of time could not conceive of any other way to organize.

Personnel Selection, Retention, and Development

After the list of World Corporations had been decided upon, and the group's structure had been set, personnel had to be selected. Tom Theobald, Head of the WCG, described the process this way:

> It was a fairly routine kind of problem during those first months of just finding about 300 people; 600 if you include the nonprofessionals worldwide. The reality is that Citibank has reorganizations regularly. None of us had ever occupied a job so long that we thought it was our terrain or turf.

Hoyle Jones, the Personnel Officer, recounted:

> We had series of meetings with the IBG (International Business Group) and NBG and figured out who was coming. People were then told what part of the organization they were going into. I don't mean to make it quite that harsh, and obviously if somebody resisted violently they weren't ordered to go. On the other hand, basically, it was announced who was going where.
>
> We took every non-U.S. citizen who was working in the NBG, with the exception of one or two in agribusiness. We tried to highlight those with multinational backgrounds. It wasn't a question of sitting down and interviewing everybody, and asking them whether they wanted to go. We told them where we wanted them to go, and in not one instance did we encounter any resistance.
>
> Whoever the Account Manager was for Mobil Oil Company in the CBG, for example, became the Mobil Oil Account Manager in the WCG. We stressed continuity. It was a fairly simple matter. If I'm an account officer and I have five names and they're all going to World Corporation Group, it's no great mystery where I'm going. Only in those instances where an account officer had a load split right down the middle, half WCG—half CBG, was there really any decision to make at all.
>
> The selection overseas was probably the most sensitive personnel issue. We were impacting rather significantly on what had been a Senof's area of responsibility.

We ended up with 320 officers: 110 in New York, 130 in Europe, and about 50 in Asia and 30 in Latin America.

To give the WCG members time to establish their identity together, and develop the necessary cohesion in working relationships, both internal and with customers, Theobald and Jones wanted to guarantee that they would be able to hold onto their personnel longer than the bank usually did. The average tenure of an account officer in the old Corporate Banking Group was about 15 months in any one assignment, and it was about 13 months in the International Banking Group. Also, part of a Louis Harris poll that had been commissioned by the WCG showed that corporate cleintele scored Citibank very low in continuity of personnel. Within the first few months, therefore, WCG established a rule that personnel transfers were not allowed to flow out of the group without a double signoff. This meant, for example, that someone could not be transferred from the WCG to the IBG in the Philippines without the approval of the senior vice presidents and the personnel officers from both groups. As one IBG officer expressed it:

It was like a vacuum cleaner. They took people, but they wouldn't release anybody. In retrospect, I think, Tom [Theobald] was right; it was to preserve the integrity of the group during the first year or so. If he started giving people back he'd be very vulnerable.

The head of the IBG points out:

We work together very well on salaries, budgets, promotions, and so on. We keep the two groups in balance, but we can do a lot more in terms of manpower development and planning. We are now programming transfers between the groups in both directions. We just made a vice-president of the WCG in Japan the Senof in Taiwan.

Besides closing the back door, WCG also instituted a policy whereby all assignments in the group were for three to five years. As an example, Jones said:

The account mix might change a little bit, but you would stay in Frankfurt for three to five years. The strategy of the bank in developing people traditionally has been one of velocity of experience. But our guys had a "Don't Touch" sign on them. In the WCG we endorse the Bank's policy; we just elongated the moves.

For some there were other drawbacks to this policy:

You can see a career path in the World Corporation Group that can take you all the way up to the executive vice-president level. It's a relatively high, narrow, and specialized career path, from account manager to the management of account managers.

Jones felt that during the first year and a half there were two key personnel problems in the WCG. First, the group began with a lop-sided skills base. People that came from a New York domestic corpo-rate banking background were strong in corporate finance but lacked international experience, and those that came from overseas often had the opposite strengths and weaknesses. Theobald, therefore, in-vested $700,000 in a four-week training program, specifically designed for the 320 professionals, on how multinational businesses operate. The second personnel priority was to develop a performance appraisal system that would assess individuals against their accomplishment on six high priorities, chosen in each case by the manager and his boss. When asked what effect these efforts have had, Jones replied:

> Our PCE [Profit Center Earnings] in 1974 were 60 percent higher than in 1973, with the same people. Projected earnings for 1975 are 47 percent higher than for 1974, with the same people, give or take a few. Clearly there has been inflation, but those are still big numbers and gains, and it's got to be related to the basic skills and techniques.

From the Perspective of a Banker in the IBG

From January 1972 until March 1973 I was head of the corporate bank in São Paulo, Brazil. This was about 50 percent of the Brazilian market. Then I went to the head office in Rio to be Administrative Officer.

The creation of the WCG caused a lot of fear and apprehension at first. The guy who took my job in São Paulo all of a sudden found that it was split apart, and he jokingly referred to himself as *los restos,* the remain-ders. There was very much an image that the WCG was going to be the green berets, the superforce. They were going to have all the sexy accounts, and we in IBG were going to have what was left over.

In European countries, like Germany and Italy, WCG was taking some-thing like 80 percent of the profits away from the Senofs. This was one of the first direct cuts at the geographical responsibility of the Senofs. There had been matrixing before; it had been coming in gradually. This was the first direct time where they said, "Senof, thou shalt matrix with somebody called the World Corporation Head. You guys shall sign off on joint budgets. He now has main responsibility for those 400 [sic] accounts in your country, and he doesn't report to you anymore, he re-ports to us." So, a Senof was naturally anxious. Our leading edge has always been geography, but organizing by customer segment or by pro-duct line begins to chip away at the geographic authority of the Senofs. The bigger the chip, the more understandably nervous was the Senof.

In Brazil, the results turned out to be not nearly as bad as we initially thought. IBG had major responsibility and promotions kept coming. World Corporations only accounted for 25 percent–30 percent of poten-tial profits, so it was not a major threat to the integrity of the Senof. We do so much business with the government sector and with large local corporations that the bulk of the profits remained with the IBG.

The problems that came up right away were characteristic, I think, of the WCG anywhere in a country with a tight money market. There is one pool of local funds, and it's not unlimited. So how do you divvy them up? There was a lot of jockeying and negotiating between the Head of the IBG Corporate Banking Department and the WCG Brazilian Department Head. They were not equal players; the IBG man was a vice-president with much more seniority, but the WCG man reported directly to New York. So these two had to decide on the allocation of available local funds: should they go to the most profitable accounts? do you separate current and time deposits in your figuring? do you base the decision on how much they are borrowing and lending?

Each country has an Asset and Liability Committee (ALCO) that meets weekly to decide on the funds allocations. In Brazil it consisted of these two officers, the Senof, the Treasurer, and me. We'd do a cash forecast for the week; this is the projected inflow of deposits, and the cash running off. Who gets it? The biggest division was between what we then called the BMG, Bank Marketing Group, and the WCG. As best we could figure, taking all time and current deposits, BMG was contributing about 70 percent and WCG accounts were 30 percent. So we said, "OK, regardless of who does what, for the next six months, we'll go 70 percent–30 percent." The principal idea was that the funds should be allocated through a central committee, based on some kind of agreed-upon formula. The WCG man, however, was lobbying for a separate long-term pool. He said, "My customers are contributing the higher share of the time deposits; therefore, I should get 30 percent of the short-term cash, but 50 percent of the long-term money." Nevertheless, in Brazil the major constraint is funds gathering, not putting the money out. There's no big money market as in London or New York, where you post your rate and you buy it. After looking at our alternatives we decided that since the major constraint was the lack of a local money market, we would allocate the money going back out in the proportion that it was contributed.

In Brazil the credit market is such that you can always put money out, and the funds allocation was always oversubscribed. There were times when a World Corporate client would need a vast amount of local money. That would throw the ratio for the week out of line. But if it was a good deal for the bank as a whole, WCG would get the money and would give back something later on. It's horse trading and it worked very well. The two main parties always came to an agreement, and the conflicts over funds allocation never went to a higher stage for resolution.

One of the reasons things went so well in Brazil was that we had set up a pilot WCG in Rio that worked within the normal IBG country context. When the World Corporation Group was formed in New York, the guy who became country head for Brazil came out of that pilot group. Everybody knew him and had confidence in him, and he knew everybody on the other side. Also, when he formed his department, many of the people just changed their boss, but didn't change their accounts.

Another reason I feel that it has been successful is that WCG started with a given stable of business. Difficulties in a matrix come when you create a new unit that has no business, and they start reaching through, trying to find it, cutting across the existing lines.

It has been a very clean, well-managed transition. The WCG Country Head is put in a very interesting position, where he is absolutely dependent on the operational and administrative support of the country [IBG] to get his job done. So, therefore, he was forced into a diplomatic stance. If he wanted telephones, stationery, or calling cards, he had to come to the Senof's Purchasing Department. You don't want to build completely parallel integrated organizations. The danger to do so is there. If a WCG country head feels that he's not getting adequate support he might try to sneak in some of his own numbers men, his own planning people. He might try and create a duplicate MIS staff and a marketing plant, but he can't very well create a discount department or a loan department. Still and all, it would be very counterproductive. We were told, "You guys cooperate. The WCG man is vulnerable as a newborn babe, but the business is damn profitable for the Institution." Even if you get mad, you don't want to screw it up.

From the Perspective of Officers in the WCG

Tony Howkins, the WCG Division Head for Latin America, added to the perspective of the Brazilian example:

In Brazil the Senof is a strong centralist; a good, well-organized manager, with a terrifically complicated job. We represent a very small part of this organization, about 30 people out of 1000. So we are absolutely dependent on the IBG and the Senof to supply us with heat, light, space, backroom services, funding, and to establish the appropriate salary brackets for local employees. In many of his branches, where there isn't enough WCG business to have our own people, our affairs are looked after by IBG people. Yet we have the responsibility for the credit and the profitability of those relationships. There's a terrific interplay.

Because Brazil has a continuously short supply of cruzeiros, one of the big matrix problems is who gets how much of the available supply. We obviously get to keep what we generate in demand deposits and, to some extent, time deposits. But then we go into the pool of funds that are generated through smaller branches which don't get the same degree of demand; or funds that are generated by the Treasury Department or through swaps of dollars into cruzeiros, or what have you. There are many mechanisms to create cruzeiros. We have to sit down and negotiate with the IBG, and the Senof is the last word on how much our share is. His stated responsibility is as the chief strategist for Citibank in the Brazilian market. If the multinational market were being attacked locally, he might decide that we ought to go a little easy and not put so much money into the

World Corporations, adopt a strategy of supporting the government's determination to build a stronger agricultural sector. Through fiat, then, he might determine that our share of funds will go down from 30 percent to 8 percent.

The other very critical matrix problem in capital-scarce countries like Brazil is that there is no readily apparent way to determine just exactly what is the value of the pool of funds available, the cost of lending. The managerial intention of the pool is to bear the incremental cost of funds so that account officers are always operating off the incremental notion of what is the value of those funds. We price according to this incremental value rather than, say, the average cost of funds, which is the way we perceive that a lot of our competition in fact operates. The value of this is that it sends more correct signals to our account officers. Determining the incremental cost of funds is a subject of potential and sometimes real friction in the ALCO meetings in some of the capital-scarce countries.

Meeting the interdependent needs of the WCG and the IBG in each of the designated countries was a process that could not be mandated by headquarters in New York. A WCG Country Head describes how he established his viability and working relationships vis-à-vis the IBG organization in the same country.

The Country Head had to realize where he was vulnerable if he was to establish his department. He was vulnerable to getting funds, people, and office space. During the first six months we had four or five major battle on these things.

In one instance, the Senof had a problem in operations and wanted a man back, who had been transferred to the World Corporation Department.

If early on in the game the Country Head said to the Senof, "Oh, yeah, we're going to be matrixing together and you need that guy; sure, go right ahead and take him back," that would have been the destruction of the matrix. He wouldn't have established a balance of power, and the negotiations stop at that point. For the first six months, the enlightened Country Head had to create this independent group, and make sure that it was clear to the Senof that if any decisions were made, I had a reporting line that I had to check back with. The balance of power came into effect because here's my Division Head backing me up. Now, the Country Head's other problem is that he knows if he brings the problems to his boss, he's an ineffective manager. So in most instances, the Country Head used the argument with the Senof, "Look, if we can't resolve it here to our mutual satisfaction, we're going to get an SVP in New York involved, and we don't want to do that. It's not to the benefit of either one of us; let's resolve it here. Two or three Country Heads didn't do this, and the Senof still exerts too much influence and the matrix is weak there.

After six months, we started the integration process. By then, we had gotten the independence necessary: we're all part of a team here, and we are a separate but equal unit. It was like the United States informing Britain that the nation was formed and we're up on our feet; now you can send representatives over and we'll receive them. But we are separate. We've got our own constitution and the reporting lines are different. So there were a series of meetings and then, I think, the spirit of one country, two teams, came.

Another thing that helped was that right after the reorganization, Theobald and Vojta, the two Group Heads, made three or four trips together to the country organizations. They had a pretty well orchestrated dog and pony show of joint presentations. Then, they would talk and answer questions together, giving the same line and reinforcing each other. I understand that when one of them goes to a country, he gets a briefing from the whole country organization including the other group. They spend time with each other's people, they represent the Institution. The whole organization has to turn out and perform.

The Country Head summarized his thoughts about why the WCG became successful so quickly.

It's a small group of people who can identify with a fairly large piece of business and work as a well-managed team with a very well-defined, well-thought-out concept. One of the main keys is the size of it; it's a manageable piece of business. The IBG is so large and complex, there's got to be more of this market management concept. I think the Senofs are now cooperating, and I see other pieces of groups coming out of the IBG, like the Merchant Banking Group came out. Matrixing is certainly working.

It is in the same vein that Vojta from the IBG quotes Theobald when defining his conception of the new role of the Senofs.

Tom has a great phrase. He says the Senof is an "integrator of sensitive processes." As a user group, Theobald sees the functional responsibilities of the Senof as the integrator of sensitive processes, seeing to it that the corporation remains balanced and managed, and true to its customers and sovereign governments. The Senof sees that the differing points of view are iterated and negotiated until you reach a point of consistency that presents the corporation as a whole. He makes that happen as the presiding godfather. Costanzo does the same thing at his level, with both the WCG and IBG reporting to him.

As a postscript, Vojta adds that since the reorganization, the Senof in Malaysia has requested twice that Malaysia be named a designated country in the WCG.

Although they have 70 World Corporations doing business there, none have headquarters in Malaysia. But what a terrific indicator of success: they want it!

From the Perspective of a National Account Manager in the NBG

There was some concern in the beginning but, on balance, nobody got too hurt. I think the principle was that everyone should be perceived as having gained. We did lose some major accounts, but then we picked up one whole department, which brought this department back to the size it was prior to the reorganization. Before, we had 30 people handling about 50 client relationships; we were doing roughly $400 million-plus in loan volume. Subsequently, we lost a substantial amoung in loan volume, but less in the number of customers. From the merger of the departments, we ended up with essentially the same loan volume. So, as a department manager looking at his asset base, span of control, number of people that he supervised, and so on, it doesn't change essentially. Another positive aspect was that we inherited companies that require more hands-on management. These companies are more risky credits, they involve more time and effort, and are more challenging to work with and to help. On the negative side, we did lose some key companies and some of the international flavor.

After any organizational change, there's always a point where you have a feeling as to whether you ended up in the right place. At the time of the restructuring there was a lot of discussion, particularly among the account managers, as to where they should go, and where they would like to go if they had a choice. Not many of them did. It all boiled down to the type of customer concentration.

What is the right side of the fence to be on, in terms of my potential advancement in the Institution? I think there was a real mix of feelings, a 50–50 split. In the WCG there's less of a challenge in terms of financing, although there's a great marketing challenge. In the NBG there's a marketing challenge, but there is certainly a higher degree and a more frequent opportunity of dealing with customers that require financial consulting and closer observation of what's going on in the company. Still, there's a feeling among most of the younger officers that they need a tour of international duty, either abroad or in the WCG.

Today there is very little interface between the WCG and the NBG. We started out with the concept that we would try to have good and frequent communications. But it really hasn't worked out because the job doesn't call for it. The portion of any one industry in each group is significantly different, so there is not any real need or motivation to do so. In some industry departments there is more cross-exposure; in mining, for example, they do share the mining engineers. But among the account managers in the two groups I don't think there is that much communication.

Future Organization

The successful creation of the World Corporation Group, together with many other major organization changes during the past two years, has led the principal executive officers of Citibank into informal discussions about whether there is any conceptual consistency of design. Will any coherent model emerge as a basis for giving structural form to future banking innovations? Does each major change create its own form, ad hoc, or can the bank find a unifying framework around which to organize its diverse strategies? Listen to three different perspectives, not necessarily representing three specific individuals, of the senior executives at Citibank.

Executive Vice-President 1

My personal view is that we ought to organize around constituent groups worldwide. The bank ought to be organized more economically and efficiently along global lines: global financial services, global consumer lines, global investment banking lines, global corporate lines, global operational lines.

In any given country we would probably integrate across those business markets much as we do now in New York. If you can penetrate the consumer market nationally, in the United States, believe me you can penetrate it anywhere. I think you need to organize around a commonality of the skills necessary, not around an integration of geographies. There's more interdependence in the consumer business in different parts of the world than there is with two different businesses, like corporate and consumer, both in the same country. We organized around geography a lot, but then we grouped around specialized industries and a lot of other things; so this would not be flying in the face of traditions. The real economy, here, would be to have central cores of industry expertise applicable any place in the world. We're not there yet, but the evolution dictates that maybe we will get there.

Executive Vice-President 2

We're big enough to tolerate a multiplicity of models. I don't think any of us mind as long as any one model doesn't get dysfunctional. The model we used for the leasing business was perceived as having gotten to be dysfunctional. This was what you might call the "independent operating company" or the "conglomerate" model. Up until last year we had a thing called Citicorp Leasing. It was run as a separate corporation out of Citicorp, the bank's holding company. It had its own operating base, its own accounting structure, its own treasury function, its own marketing structure, and so on. It ran leasing domestically and internationally. There were some tenuous linkages to the existing structure domestically, and overseas the Senof was supposed to have a joint signoff on the business plans, and so on. But, in fact, it was very weak. Over time, the tensions got so bad that last year, as we created the Merchant Banking business, we broke the leasing organizations up into a domestic piece that went into the National Banking Group and the international part which went to the IBG. In effect, we are in the process of reintegrating and reconsolidating it.

Now we broke out the Merchant Banking business in the same way as we did with leasing and I sense the same kinds of tensions are going to be repeated. My perception is that this carve-out, independent model isn't going to suit us over time, but it may be right as the first step in a model. That's been the experience with leasing. Nobody knew how to do leasing, [so] we split it up from the rest of the bank. We got guys in from the outside and it was hellish and traumatic, but we did build a hell of a leasing company as a result. Now it's integrated in the bank. I think we all know enough about it now so that it won't be smothered. It will be kept in the family as a legitimately differentiated line of business. Whereas, if you hadn't spun it out, it probably would never have gotten off the ground, except for a few places here and there. That's the great advantage of this independent operating model. The trade-off is the organizational tension that it introduces, which is considerable. When you matrix a market or a product line, you do it for a while, until you get the degree of differentiation that you want, and then at some point you may integrate it. We do a lot of matrixing.

Executive Vice-President 3

The matrix organization that we are using is really a way to promote diversification in the business, by market, product, or whatever. You're trying to mine that segment of the business more deeply by putting professionals, narrowly focused, at that point. We are better off in the World Corporation market today because we have a bunch of guys dedicated solely to that market.

We have another variation of that in the consumer market. John Reed has got corporate responsibilities for it, but he has no line organization at all overseas. He works totally through the IBG structure, abroad. He operates off the IBG infrastructure the way WCG does. And, as in the WCG, in the last few weeks the bank has chosen a couple of countries to be designated consumer countries for this purpose. In New York, the consumer business is run through Jim Farley and the New York Banking Group. Reed supervises directly the non-New York State consumer vehicles, in the United States. It's a real hodgepodge at this point. We haven't gone to the extent of announcing a matrix organization, like WCG's. We have not said that John Reed runs the consumer business worldwide, in a line sense. He operates as an extension of the Chairman's and President's offices, working with the line groups in the organization. But, essentially, it's the same model. Over time, it will probably become exactly the same, after we fuss around with it a bit more.

The Investment Management Group offers a third example of this model. John Heilshorn has international responsibilities for a discrete product line, but he uses the IBG infrastructure as the WCG does. He's had a smooth time getting directly involved internationally because he brings demonstrably specialized skills which are absent in many of the international markets.

From these examples, a model emerges. You integrate the infrastructure along geographic lines and you differentiate the marketing along business lines. I have a conception of it, but I'm not lobbying for it at this point. We're sensitive to the pace of change and how much we can do at one time. Also, we're in a difficult business environment. Still, if you asked now for a senior management vote, you would probably get nine votes to three in favor of this model.

Moral: Matrix is a verb.

NOTES

1.   Peter F. Drucker, 1974. *Management: tasks, responsibilities, practices.* New York: Harper & Row, p. 598.

2.   The Harvard Multinational Enterprise Project is a large-scale study that has been going on since 1965 under the direction of Raymond Vernon. The 60 percent figure comes from one of the study's volumes. John M. Stopford and Louis T. Wells, Jr., *Managing the multinational enterprise.* New York: Basic Books, 1972. p.21.

3.   Gilbert H. Clee and Wilber M. Sachtzen, 1964. Organizing a worldwide business. *Harvard Business Review* (Nov.–Dec.), p. 65.

4.   Stopford and Wells report that when foreign sales reach 40 percent of the total, most firms turn to some form of direct area coordination, p. 64.

5.   For a detailed account of these changes, see Stanley M. Davis's case series. *First National City Bank: Multinational Corporate Banking (A), (B), (C), and (D).* Harvard Business School, International Case Clearing House, 4–467–079, –080, –081, and –082.

9
FUTURE OF THE MATRIX

THE MATRIX LIFE CYCLE AND BEYOND

In a book devoted to a new form or practice there is a natural tendency to view that new item as the ultimate of its kind. Here, there is a danger of viewing the matrix as though it were a final and complete kind of management and organization. When managers are fully involved in and committed to a successful matrix, they are not likely to ask what is next, what lies beyond the matrix. In fact, however, the matrix may not be a final form for many organizations.

In Chapter 3 we said that the matrix had a life cycle, with definite phases: from Phase I, the traditional pyramid, with its unity of command, and the three conditions which can make the pyramid inadequate; to Phase II, a temporary overlay of coordinating mechanisms such as project teams; to Phase III, a permanent overlay of the secondary dimension; to Phase IV, a mature matrix that balanced the two organizing dimensions equally. While most organizations will stabilize at Phase IV, for others there also is or will be a Phase V.

If the passage from Phase I to later phases is from unity of command to duality of command, then the passage to Phase V is from duality back to unity. But, there are several significant differences between the early and late phases in the matrix cycle. What frequently happens during Phase V is that the basic organizing dimensions are still considered equally important, but now some are more equal than others. Occasionally this may mean a return to the initial pyramid and ranking. We have found more often, however, that this final phase is a continued rotation of the organizing axes to the point where the secondary or overlay dimension becomes paramount, and the earlier predominant focus has become the supportive and the less dominant one.

Unless we were to speak of a complete failure with the matrix, in which case change would be like a regression to an earlier form, an organization reaching Phase V is clearly distinguishable.

In some successful cases, firms will settle into and happily institutionalize the structure described in Phase IV. Other organizations will find that the evolved balance continues to evolve so that the second, or added, dimension has gained primacy. In a functional structure (Phase I) that has added the product dimension (Phase II and/or III), for example, the balance between the two might (Phase IV) yield to a product division design (Phase V). Or a firm organized by product groups (I), and coordinated by geographic sectors (II, III, or IV) might shift to a geographic form that integrates product responsibility through other mechanisms (V). Hanging onto an exact structural balance is not a key to a healthy matrix. Because the start of a matrix is often difficult and unstable, the key in Phase V is to maintain the dual perspective through the other elements—behavior, systems, and culture—while the structure is allowed to cycle through to a new, more stable arrangement.

In its use of the matrix, an organization in Phase V has found how to derive the benefits of simultaneously organizing and managing paradoxical needs, without resorting to dual or matrix structures. You will recall that in Chapter 2 we said a matrix organization included matrix behavior, matrix systems, and a matrix culture, as well as a matrix structure. It seems that after years of working with a matrix, some organizations find that they no longer need the paradoxical architecture of the matrix structure in order to accomplish their goals. Rather, they go to the simpler pyramid for their structural form, while at the same time they maintain the dual or multiple perspective in their managerial behavior, in their information processing, and in their culture of their firms.

A helpful perspective for organizations that have had moderate success with the matrix but who remain uncomfortable with the complexity in the structure may be to view a power-balanced structure as a vehicle for the transition from one stable start to another. The formal design, then, is acknowledged as awkward or difficult, but also useful and necessary. It is workable because it is the framework for transition:

$$\text{from } \frac{A}{B} \longrightarrow \text{to A and B} \longrightarrow \text{to } \frac{B}{A}$$

Firms that have viewed the process of change this way, generally described a successful experience. They have a greater chance in the future of maintaining the matrix perspective and culture with the simpler formal design. Organizations that view the process as an unworkable design, that is thrown out in order to return to the simpler form, here simply exchanged one horn of their dilemma for the other. They have gone from either to or. The distinction we are making between successful transition and a scrapping failure is descriptive, not prescriptive. This is what we have seen happen. How to manage the success depends on nonstructural elements in the matrix.

This interpretation suggests that the matrix is not likely to become the dominant feature in the *structure* of American organizations. Its utility, according to this line of logic, is more likely to be in helping organizations become more flexible in their abilities to respond to environmental pressures. Structures are intended to channel people's behavior in desired ways. Like laws, they are strongest when they are not invoked or tested. To the extent that managers behave effectively, they have little need to bump up against formal structures and reporting walls. In the traditional pyramids, managers were always bumping—the structure was either centralized, and there wasn't enough freedom, or they were decentralized, and there wasn't enough control.

Organizations with a mature matrix therefore appear to follow one of two paths, and the extent to which the structural framework survives depends on which path an organization takes. One is to stay in Phase IV, maintaining dual command, shared use of human resources, and an enriched information-processing capacity. The other is to evolve into what we call Phase V, maintaining matrix behavior, matrix systems, and a matrix style or culture, but without using the matrix's structural form. In these instances, what began as a secondary overlay more often becomes the primary dimension, rather than a return to the pecking order in Phase I. Some organizations, of course, tear down the matrix entirely and revert to the traditional forms, practices, and managerial behavior of the first phase, as we described under pathologies in Chapter 6.

The distinction between a pathological breakdown and an evolutionary rotation, where the matrix is a transitional form, is a matter of interpretation. As we observe the change in these organizations we may ask, was the matrix "thrown out" or did the firm "grow beyond it?" The distinction is more than academic. So long as the environmental pressures that initially propelled an organization into a matrix remain, the original inadequacies of the pyramid form will reappear if the matrix is actually abandoned. Our observations suggest that this would be fairly evident in three to six months, painfully obvious within one to one and a half years, and back to the same kinds of discussions we find in Phase I shortly thereafter.

Because the structural element of the matrix is so "fiendishly difficult" to many, we observe organizations trying to shed the form yet maintain the substance. Our diagnosis is that it can be done successfully only in organizations in which appropriate matrix behavior is so internalized by all relevant members that no one notices the structural shift. Even then, however, we anticipate that through the years the structural imbalances will increase.

WHERE WE STAND ON THE LEARNING CURVE

We began this book with the comment that not too many years ago few managers in our classrooms had heard of matrix organization and that today nearly half of them raise their hand when asked how many work in a matrix. Objectively, this self-reporting is inaccurate. What is relevant, however, is the perception itself. Like Molière's gentleman who was surprised to find he had been speaking prose all his life, many managers are finding that they have been "matrixing" all along. The word is jargon, but the grammar connotes people's behavior more than organizations' form. The unrealistically high self-reporting also demonstrates an increasing comfort and familiarity with the idea among a very large body of executives.

Our major purposes have been to broaden traditional treatments of matrix by (1) demonstrating its applicability in many diverse settings and (2) suggesting ways to change a radical conception into a familiar and legitimate design. Matrix seems to have spread in spite of itself. It is complex and difficult; it requires human flexibility in order to provide organizational flexibility. But the reverse is also true. For these reasons, we believe, many managers shied away. The academic literature until now limited the utility of matrix to high technology, project organizations. We have shown that in both organization theory and application the matrix has a much broader place. Behavioral descriptions were replete with words like "tension," "conflict," and "confusion." For many, the matrix was not pleasant but it seemed to improve performance. Success gave it legitimacy, and as the concept spread, familiarity seemed to reduce the resistance.

Matrix gained active currency in the space age of the late 1960s. In fact, for a while in the early 1970s it almost seemed like a fad. Organizations that should never have used it experimented with the form. It was in danger of becoming another hot item from the behavioral science grab bag for business. Where this occurred, the results were usually disastrous. Some thought that an organization which played with the matrix might easily get burned. Despite many misadventures, however, matrix gained respectability. What was necessary was made desirable.

As the concept of matrix matures, interpretations of it are changing too. More people feel "it isn't that new or different after all." This is a sign of acceptance more than it is a sign of change or moderation of the design. But remember, managers lived with the happy fiction of dotted lines for generations. Formal acknowledgment of a second reporting line was always unacceptable, but it was necessary even if it wasn't kosher. So the dotted line came into being. As subordinates, people had second bosses, even if they weren't their real bosses. The intrusions of the second boss were resented. They weren't legitimate, even if they were critical. As a boss, people had to share their people with a protected interloper. The results were a kind of executive *ménage à trois*: A triangular relationship in which the 2-boss manager has two arrangements; one legitimate and another that exists but is not granted equal privileges.

Now, a long time later, what was true in fact is finally passing into formal and established theory and practice. The "copy to" boss has gained equal rights. The need for the division controller, for example, to have lines to both the division general manager and the corporate controller was always acknowledged. But one always got the nod. The matrix ensures that both functional and business-market needs are equally represented. Where it is not necessary for equal and simultaneous representation, the single chain and the dotted line are preferable. But where balance is called for by environmental pressures, organizations will be impelled to the matrix despite their initial misgivings. It is this, and not the increased pace of change, that is popularizing the matrix.

We observe that more organizations are feeling the pressure to respond to two or more critical sectors simultaneously; to organize by function *and* by product, by service *and* by market area at the same time. There is an increasing pressure to improve information-processing capacity, and the recent technological advances make multiple matrix systems feasible. Finally, it is clear that there is an increased sense of the scarcity of all resources, and hence pressures for achieving economies of scale. These were the necessary and sufficient conditions we described for the emergence of matrix organizations. Because these conditions are increasingly prevalent, we feel that more organizations will be forced to consider the matrix form.

Each organization that turns to the matrix does so with a larger and more varied number of predecessors who have charted the way. Despite our belief that matrix must be "grown" from within, the examples of wider applicability must nevertheless suggest that we are dealing less and less with an experiment and more and more with a mature formulation in organization design. Familiarity, here, reduces fear. As more organizations travel up the matrix-learning curve, the curve itself becomes an easier one to climb. Similarly, as more managers gain experience operating in matrix organizations, they are bound to spread this experience as some of them move into other organizations on their career journeys.

When pioneers experiment with new forms, the costs are high and there are usually many casualties. This was true for both organizations and individuals in the case of the matrix. As the matrix has become a more familiar alternative, however, the costs and pressures have been reduced. Today, we believe that the concept is no longer a radical one, the understanding of the design is widespread, and the economic and social benefits have increased.

People in the Middle Ages had a very clear view of the world order. Galileo changed that. Newton changed the view of universal order once more, and Einstein did the same in a later age. In each period there was certainty of the logic and correctness of the structure of the universe. Each period lasted until a new formulation posed a previously unthinkable question. After varying periods of resistance or adjustment, people become comfortable with a new formulation and in each instance assume it to be the final word.

The organization of large numbers of people to accomplish uncertain, complex, and interdependent tasks is currently nowhere as susceptible to the exactness in calculation as the physical world. And there are those who would say that to compare the world of physics and the world of organization is to compare the sacred with the profane. But the process of acceptability and then increased applicability of new formulations is similar, even if rather more humble. We believe, therefore, that in the future matrix will become almost commonplace and that managers will speak less of its difficulties, and assume more of its advantages.

A PRODUCTIVE AND HUMANE DESIGN

In the beginning of this book we said that the models for organizing business often come from other institutions. The same pattern and dilemma as in the matrix, for example, was familiar to those who wrote the United States Constitution. They recognized the tendency for balanced power to revert to unitary and absolute form, and therefore checks had to be established and defended formally in order to maintain liberty. The best behavior, by the best intended people, will not be sufficient. The purpose of polity is to protect liberty while maintaining order. We have argued that the matrix principle applied to economic organization can accomplish similar objectives: the matrix can be both more productive and more humane than the unitary models.

One of us recalls delivering a lecture on the matrix to a mixed audience of managers and academics at Berkeley. One of the skeptics in the audience asked, "How can you say that the matrix increases personal freedom as well as organizational productivity if, in an organization of 50,000, only a few thousand at the middle and top are directly in the matrix?" Our answer is that our government operates as a balance of legislative, executive, and judicial bodies in the local, state, and national jurisdictions. Although the average person is not directly involved in this design on a daily basis, it is a strongly held belief that our lives are better for it nevertheless. Similarly, we acknowledge that most members of large organizations that use the matrix will not be aware of it, or its benefits, in daily terms. We suggest, however, that the matrix provides a more flexible design than does the unitary model of organization and command, and it also offers the potential for increasing personal freedom in a more productive workplace.

GLOSSARY

Balance of Power An equilibrium derived from the acknowledgment that there is no one best way to organize, that it is workable to organize along two dimensions (function, product, area, etc.) simultaneously, and that the benefits of each are not mutually exclusive.

Dual Command The principle of shared power over a common subordinate.

Matrix Diamond The elementary structural form of a matrix organization; in contrast to the pyramid, which is the form for traditional organization design.

Matrix Manager A manager who shares power with another manager over a common subordinate.

2-Boss Manager A manager with two reporting lines, equally responsible to two different superiors in the organization.

Phases - Life Cycle Five identifiable periods in the evolution of matrix management and organization from entry to exit.

SELECTED ANNOTATED MATRIX ORGANIZATION BIBLIOGRAPHY

Ansoff, H.I., and R.G. Brandenburg, 1971. A language for organization design: Part I and II. *Management Science* **17** (12): (August) B-705-31.
An excellent review of the project management literature. Deals extensively with the conflict faced by professionals involved in project work.
Authors provide a complex but general language for the logical design of four different structural forms, one of which is the "adaptive" or matrix form.

Argyris, Chris, 1967. Today's problems with tomorrow's organizations. *Journal of Management Studies* (February): 31-55.
An empirical study of nine British matrix organizations. The study is positive about the structure, but demonstrates how implementation has been unsuccessful on account of traditional management behavioral styles.

Butler, Arthur G., 1973. Project management: a study in organizational conflict. *Academy of Management Journal* **16** (1): (March) 84-101.
An excellent review of the project management literature. Deals extensively with the conflict faced by professionals involved in project work.

Cleland, David I., and William R. King, 1968. *Systems analysis and project management.* New York: McGraw-Hill.
One of the best and most thorough books explaining project management and locating it in the larger setting of systems and organization theory.

Davis, Stanley M., 1974. Two models of organization: unity of command versus balance of power, *Sloan Management Review,* (Fall): 29–40.

This article spells out the theoretical foundations for the model used in this book.

Davis, Stanley M., and Paul R. Lawrence, 1977. *Harvard Business Review,* Nov.-Dec.
Chapter 6 of this book.

Delbecq, Andre L., Fremont A. Shull, Alan C. Filey, and Andrew J. Grimes, 1969. Matrix organization: a conceptual guide to organizational variation. *Wisconsin Business Papers* No. 2 (September). Madison, Wisconsin: University of Wisconsin Bureau of Business Research and Service.
Paper presents key dimensions of a matrix organization model developed by the *Matrix Management Project Group* at the University of Wisconsin. This NASA-funded activity was carried on for several years and then dissolved.

Galbraith, Jay R., 1971. Matrix organization design. *Business Horizons* (February): 29-40.
Through treating a fictitious case, author describes the decisions involved in the process of adding a product orientation to a functional organization until an appropriate balance is reached. Article delimits the boundaries of matrix organization.

Goggin, William C., 1974. How the multidimensional structure works at Dow Corning. *Harvard Business Review* (Jan.-Feb.): 54-65.
A case description of how Dow Corning expanded a matrix form of organization into one that added an area dimension to the product and function dimensions, plus a fourth dimension to consider organizational evolution.

Grinnell, Sherman K., and Howard P. Apple, 1975. When two bosses are better than one. *Machine Design* (January 9): 84-87.
A very brief, but very good, set of practical guidelines for when to use a matrix organization and how to make it work.

Kingdon, Donald Ralph, 1973. *Matrix organization: managing information technologies.* London: Tavistock.
Attacking the inappropriateness of hierarchical structures in turbulent environments, the author proposes a sociotechnical approach to matrix organization. The approach concentrates on work groups and has a strong organization development orientation.

Lawrence, Paul R., and Harvey Kolodny, and Stanley M. Davis, 1977. The human side of the matrix, *Organizational Dynamics,* Summer.
Discusses the key roles in the matrix. Adapted from Chapters 3 and 4 of this book.

Likert, Rensis, 1975. Improving cost performance with cross-functional teams. *Conference Board Record* (September): 51-59.
Author advocates the use of matrix organization to sustain cross-functional teams which, while productive, tend not to endure because of the power in the functional structure.

Marquis, Donald G., 1969. A project team + PERT = success. Or does it? *Innovation* (5): 26-33.
Relevant report of an empirical study of 38 large projects which were organized in different ways. The merits of projects versus function versus matrix are considered.

Mee, John F., 1964. Ideational items: matrix organization. *Business Horizons* 7 (Summer): 70-72.
Very brief note that provided wide dissemination for the term "matrix" referring to it as a "web of relationships."

Sayles, Leonard R., and Margaret K. Chandler, 1971. *Managing large systems: organizations for the future.* New York: Harper & Row.
Contains several chapters which compare project and functional forms and describe the project manager's role in large projects embedded in NASA's structural framework.

Sayles, Leonard R., 1976. Matrix management: the structure with a future. *Organizational Dynamics,* Autumn, pp. 2-17.
Holds that many organizations have adopted the matrix form without the name, and identifies five different types.

Shull, Fremont A., Andre L. Delbecq, and L.L. Cummings, 1970. *Organizational decision making.* New York: McGraw-Hill, pp. 187-208.
Contains a section which describes the matrix model of the Matrix Management Project Group. (See Delbecq *et al.*)

Smith, Robert A., 1966. The matrix organization form: a social concept for enterprise effectiveness. NASA, George C. Marshall Space Flight Center, Management Development Office. (November 15).
An optimistic statement about matrix organization's capacity to override the dysfunctions of bureaucratic structure and substitute a form better able to respond to the needs of both the task and the individual in the organization.